VERMICOMPOSTING

VERMICOMPOSTING
PRINCIPLES, PRACTICE AND BENEFITS

Ted Town

Molly Jahn

Rupert Thatcher

Kruger Brentt
Publishers
2023

Kruger Brentt Publishers UK. LTD.
Company Number 9728962

Regd. Office: 68 St Margarets Road, Edgware, Middlesex HA8 9UU

© 2023 AUTHORS
ISBN: 9781787150461

For information on all our publications visit our website at http://krugerbrentt.com/

PREFACE

Vermicompost derives from the Latin word vermi for worm, as it is essentially worm castings. During this biological process, earthworms are used to transform organic waste into a nutrient-rich soil that acts as a slow-release organic fertilizer. Earthworms are valued by farmers because, in addition to aerating the soil, they digest organic matter and produce castings that are a valuable source of humus. Vermicomposting, or worm composting is a simple technology that takes advantage of this to convert biodegradable waste into organic manure with the help of earthworms (the red worm Eisenia foetida) with no pile turning, no smell, and fast production of compost. The earthworms are bred in a mix of cow dung, soil, and agricultural residues or predecomposed leaf-litter. The whole mass is converted into casts or vermicompost, which can be used as a fertilizer on all types of plants in vegetable beds, landscaping areas, or lawns.

Worms are so effective at processing organic waste that they can digest almost half their own weight in debris every day. Vermicomposting is a simple composting process that takes advantage of what earthworms do naturally, but confines the worms to bins making it easier for farmers to feed them and to harvest their nutrient-rich compost. Since all worms digest organic matter, in principle, any type of worm can be used; however, not all are equally well adapted to living in bins since some worms prefer to live deep in the soil while others are better adapted to living closer to the surface. The red worm (Eisenia foetida) is ideal for vermicomposting because its natural habitat is close to the surface and it is accustomed to a diet rich in organic matter, this makes it ideally suited to digesting kitchen scraps and to living in bins.

It is a great sustainable alternative to use as fertilizer and a fantastic way to promote sustainable agriculture in local farming. Vermicompost has a plethora of benefits, most importantly as it acts as biofertilizers, restores soil nutrients, stabilizes soil, and enhances soil fertility at a long-term period; it attends to social issues and recycles waste; and it is shown to be a profitable enterprise as a circular economy.

The present book contains eleventh chapters covering all related disciplines. These chapters include Introduction to Vermicomposting, Compost: The Black Gold, Vermiculture Biotechnology, Factors Affecting the Composting Process, Manures and Composts, Enrichment of Organic Compost, Vermiwash: A Potential Tool for Sustainable Agriculture, Vermicomposting of Agriculture Waste, Organic Waste Management by Earthworm, Role of Vermicomposting in Improvement of Soil Nutrients and Agricultural Crops, Trouble Shooting, This book elucidates the concepts and innovative models around prospective developments with respect to vermicomposting. Related terminology is given at the end for ready reference. It is believed that this book would be useful to all those who are interested in vermicomposting including industrial organizations, farmers, gardeners' researchers, teachers as well as students and scholars.

We are grateful to all those persons as well as various books, manuals, periodicals, magazines, journals etc. that helped in the preparation of this book. In spite of the best efforts, it is possible that some errors may have occurred into the compilation and editing of the book. Further queries, constructive suggestions and criticisms for the improvement of the book are always welcome and shall be thankfully acknowledged.

Ted Town

Molly Jahn

Rupert Thatcher

CONTENTS

1

INTRODUCTION TO VERMICOMPOSTING

1.1 INTRODUCTION

Rapid industrialization, urbanization, and the ever-increasing population generate voluminous solid wastes. In recent years, disposal of organic wastes has caused serious environmental hazards and economic problems. Burning of organic wastes contributes tremendously to environmental pollution thus, leading to polluted air, water and land. This process also releases large amounts of carbon dioxide in the atmosphere, a main contributor to global warming together with dust particles. Burning also destroys the soil organic matter content, kills the microbial population and affects the physical properties of the soil. A substantial portion of this solid organic waste is non-toxic. On one hand tropical soils are deficient in all necessary plant nutrients and on the other hand huge quantities of such nutrients available in solid organic wastes. Treatment of solid organic wastes has therefore become an essential part of the waste management programmes almost all over the world. Existing methods to its treatment and disposal are rather expensive. Vermicomposting technology is one of the best options available at present for the treatment of organic wastes. The term vermicomposting is coined from the Latin word 'Vermis' meaning the 'worms'. Vermicomposting refers to natural bioconversion of biodegradable garbage into high quality manure with the help of earthworms.

Earthworms play a key role in soil biology; they serve as versatile natural bioreactors to harness energy and destroy soil pathogens by feeding voraciously on all biodegradable refuse. They are nature's way of recycling organic nutrients from dead tissues back to living organisms. They have faithfully done their part to keep this cycle of life continuously moving for more than 20 million years. Ancient civilizations including Greece and Egypt were the pioneers in recognizing the role of these worms played in soil. Earthworms were considered as "sacred" by the Egyptian Pharaoh, Cleopatra who recognized the role played by worms in fertilizing the Nile Valley croplands. The role of earthworms in waste stabilization has been known for many years, especially in Southeast Asian and European countries. Attracted by the nature and activities of burrowing earthworms in soil, Charles Darwin carried out studies on the significance of their activities for over 39 years. He wrote this about these tiny creatures," It may be doubted whether there are many other animals in the world which have played so important a part in the history of the world." He also called them as unheralded soldiers of the land, in his last and final book "The Formation of Vegetable Mould, Through the Action of Worms, With Observations of their Habits" which reported how these organisms feed and convert organic materials (Darwin, C.,1881).

From then on, studies have been carried out to find out their role in maintaining the soil fertility and also in the degradation of the organic matter present in the soil. These works also included investigations into the possibility of utilizing earthworms for the breakdown of organic wastes such as animal wastes, vegetable wastes and municipal solid wastes (MSW). Earthworms convert a portion of the organic matter into worm biomass and respiration products, and expel the remaining partially stabilized matter as discrete material (castings). In this process, earthworms and the microorganisms act symbiotically to accelerate the decomposition of organic matter. The driving forces behind the introduction of vermiculture and other reuse processes, is the global recognition of the need to recover organic material and return this to the natural cycle. Vermicomposting is generally defined as the aerobic decomposition of organic residues by exploiting the optimum biological activity of earthworms and micro-organisms. The process depends upon the earthworms to fragment, mix and promote microbial activity in the organic waste material. The earthworms ingest organic solids and convert a portion of it into earthworm biomass and respiration products and egest peat like material termed as vermicompost. As compared to the thermal composting, vermicomposting generates a product with lower mass, high humus content, processing time is lower, phytotoxicity is less likely, fertilizer value is usually greater, and an additional product (earthworms) which can have other uses is produced.

Organic waste materials which are biodegradable in nature may be used as substrates for the vermicomposting process, provided that the materials do not contain anything harmful to earthworms. For example, the byproducts of agro-industrial processing offer potential opportunities to be used as substrates for the earthworms and microorganisms. The agroindustrial wastes are huge source of plant nutrients and their disposal means the ultimate loss of the resourceful material. At present, these wastes are either grossly underutilized or completely unutilized due to in situ burning in the fields or land disposal to the surrounding areas. These individually and cumulatively agro-industrial wastes could effectively be tapped for resource recovery through vermicomposting technology for use in sustainable land restoration practices.

Some of the major agro-industrial wastes explored for vermicomposting are shown in Table.1

Table.1: Details of agro-industrial organic wastes for vermicomposting

No.	Sources	Types of wastes generated
1	Agricultural	Rice husk, cereal residues, wheat bran, millct straw etc.
	Food processing waste	Canning industry waste, breweries waste, dairy industry waste, sugar industry waste press mud and trash, wine industry waste, oil industry waste-non edible oil seed cae, coffee pulp, cotton waste etc.
2	Wood processing waste	Wood chips, wood shavings, saw dust
3	Other industrial wastes	Fermentation waste, paper and cellulosic waste, tannery waste
4	Local organic products	Coco fiber dust, tea wastes, rice hulls etc.
5	Fruits and vegetable processing waste	Peels, rinds and unsed pulp of fruits and vegetables

1.2 EARTHWORMS

The earthworm is a segmented invertebrate. Its body holds its tube-shape because it is full of a liquid called coelomic fluid found between the body wall and the alimentary canal. Earthworm has a long, cylindrical body with a pointed head. In some species the posterior end is slightly flattened, while in others the body is cylindrical throughout. Rings that surround the moist, soft body allow the earthworm to twist and turn, especially since it has no backbone. With no true legs, bristles (setae) on the body move back and forth, allowing the earthworm to crawl. Earthworm breathes through its body surface. Food is ingested through the mouth into a bag like structure referred to as crop. In some species a distinct crop is absent. Later the food passes through the gizzard, where ingested stones grind it up. After passing through the intestine for digestion, what's left is eliminated as castings.

The earthworm is a segmented invertebrate. Its body holds its tube-shape because it is full of a liquid called coelomic fluid found between the body wall and the alimentary canal. Earthworm has a long, cylindrical body with a pointed head (Padashetty, S. & Jadesh, M., 2014) (Fig.2). In some species the posterior end is slightly flattened, while in others the body is cylindrical throughout. Rings that surround the moist, soft body allow the earthworm to twist and turn, especially since it has no backbone. With no true legs, bristles (setae) on the body move back and forth, allowing the earthworm to crawl. Earthworm breathes through its body surface. Food is ingested through the mouth into a bag like structure referred to as crop. In some species a distinct crop is absent. Later the food passes through the gizzard, where ingested stones grind it up. After passing through the intestine for digestion, what's left is eliminated as castings.

- Earthworms are popularly known as the "*farmer's friend*" or "*nature's plowman*".
- Earthworms belong to:

    Phylum - Annelida

    Class – Chaetopoda

    Order - Oligochaeta

- They are the first group of multicellular eucoelomate invertebrates to have succeeded to inhabit terrestrial environment.
- They are hermaphrodites, both male and female reproductive organs are present in every single earthworm but self-fertilization does not generally occur.
- There are about 3000 species of earthworms in the world which are adapted to a range of environment, in tropical to temperate region.
- About 418 species of earthworms present in India.
- India harbors about 11.1% of the global earthworm diversity.
- The Indian earthworm fauna is predominantly composed of native species, which constitute about 89% of total earthworm diversity in the country.
- Around 45 exotic peregrine worms have been introduced into India.
- The number of segments from the peristomium to the clitellum and the number of segments which make up the clitellum are species specific in earthworms.

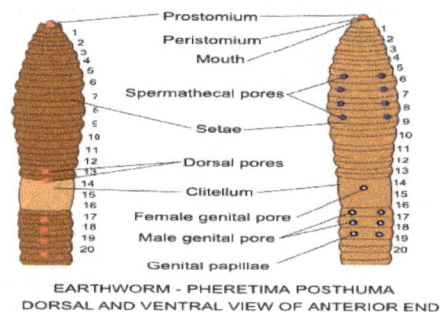

Prostomium
Peristomium
Mouth
Spermathecal pores
Setae
Dorsal pores
Clitellum
Female genital pore
Male genital pore
Genital papillae

EARTHWORM - PHERETIMA POSTHUMA
DORSAL AND VENTRAL VIEW OF ANTERIOR END

Fig. 1 Earthworm

1.3 FACTORS AFFECTING DISTRIBUTION

The distribution of earthworms in soil is affected by physical and chemical characters of the soil, such as temperature, pH, moisture, organic matter and soil texture.

1.3.1 Temperature:

The activity, metabolism, growth, respiration and reproduction of earthworms are all influenced greatly by temperature.

1.3.2 pH

pH is a vital factor that determines the distribution of earthworms as they are sensitive to the hydrogen ion concentration. pH and factors related to pH influence the distribution and abundance of earthworms in soil. Several workers have stated that most species of earthworms prefer soils with a neutral pH. There is a significant positive correlation between pH and the seasonal abundance of juveniles and young adults.

1.3.3 Moisture

Prevention of water loss is a major factor in earthworm survival as water constitutes 75-90% of the body weight of earthworms. However, they have considerable ability to survive adverse moisture conditions, either by moving to a region with more moisture or by means of aestivation. Availability of soil moisture determines earthworm activity as earthworm species have different moisture requirements in different regions of the world. Soil moisture also influences the number and biomass of earthworms.

1.3.4 Organic Matter

The distribution of earthworms is greatly influenced by the distribution of organic matter. Soils that are poor in organic matter do not usually support large numbers of earthworms. Several workers have reported a strong positive correlation between earthworm number and biomass and the organic matter content of the soil (Ismail,1997).

1.3.4 Soil Texture

Soil texture influences earthworm populations due to its effect on other properties, such as soil moisture relationships, nutrient status and cation exchange capacity, all of which have important influences on earthworm populations.

1.4 ANATOMICAL CHARACTERISTICS OF EARTHWORMS

- Bilateral Symmetry: If you cut an earthworm down the centre, you would find that the left and the right sides of its body are identical or symmetrical.

- Locomotion: They crawl using circular and longitudinal muscles which are located under the epidermis.

- Except for the first and last segment, all the other segments have eight setae located around each segment. The setae look like small bristles sticking out of the earthworm's skin. The bristle-like setae anchor the segments as they crawl.

- As the earthworm tunnels through the soil, it excretes mucus from its body. This mucus reacts with the soil of the tunnel walls and forms a type of cement which makes the tunnel walls stable.

- Respiratory System: Earthworms do not have a well-developed respiratory system. Gases are exchanged through the moist skin and capillaries, where the oxygen is picked up by the hemoglobin dissolved in the blood plasma and carbon dioxide is released.

◉ Circulatory System: The circulatory system is fully closed. One large blood vessel runs the length of the body, immediately beside the gut.

◉ Two to five pairs of muscular blood vessels extend from the central vessel and function as hearts to drive the circulatory system.

◉ Brain & nervous system: The earthworm brain is actually a fused pair of nerve ganglia, mostly located in the third segment, and the nerve fibers that run the length of the body, around the gut.

◉ Reproductive system: The clitellum is a swelling of the skin and can only be seen in earthworms that are ready to reproduce.

◉ It may be white, orange-red or reddish-brown in colour. Earthworms are ready to mate when their clitellum is orange.

◉ Most of the material secreted to form earthworm cocoons is produced within the clitellum.

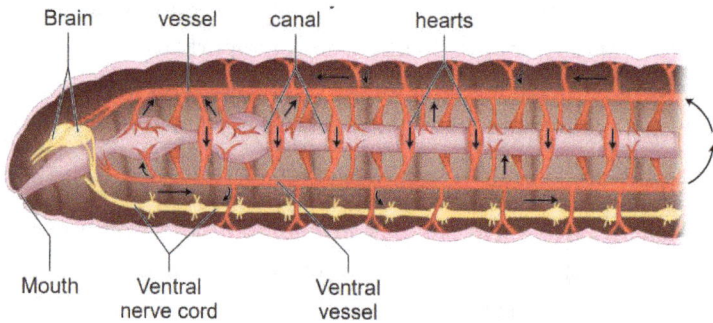

Fig. 2: *Lampitomauritii*: Circulatory system and Nervous System

◉ At the time of egg laying, the clitellum is transformed into hard, girdle like capsule called cocoon.

◉ Shedding of cocoon ranges from 1 to 5, only a few of them survive and hatch.

◉ The formation of cocoons takes a period of 50-60 days.

◉ Normally, the average life span of earthworms varies with species ranging from 1 to 10 years.

◉ Earthworms are long, narrow, cylindrical, and segmented with a glistening dark brown body covered with delicate cuticle.

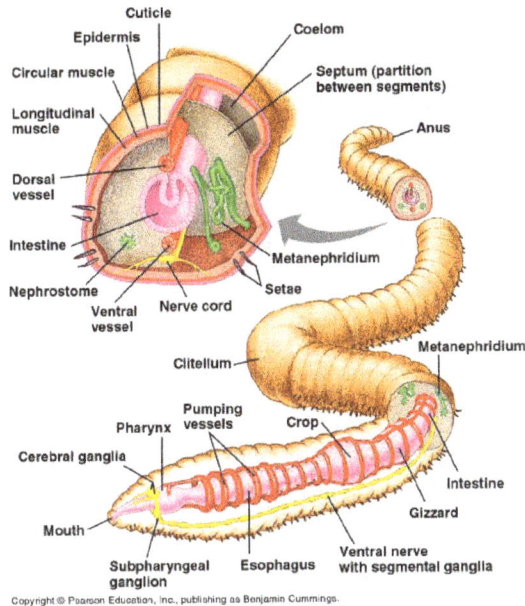

Copyright © Pearson Education, Inc., publishing as Benjamin Cummings.

- The gut of earthworm is a straight tube starting from mouth followed by a muscular pharynx, oesophagus, thin walled crop, muscular gizzard, foregut, midgut, hindgut, associated digestive glands and ending with anus.

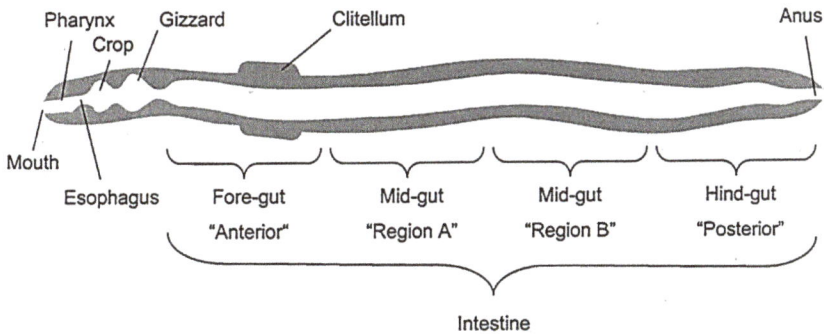

1.5 TYPES OF EARTHWORMS

- Three ecological categories of earthworms - epigeics, endogeics and anecics.

- The anecics and endogeics are known as 'soil ecosystem engineers' and their impact on soils is great and may influence properties and processes at the ecosystem level.

- The functional role of epigeics is primarily that of 'litter transformers', like other litter invertebrates.

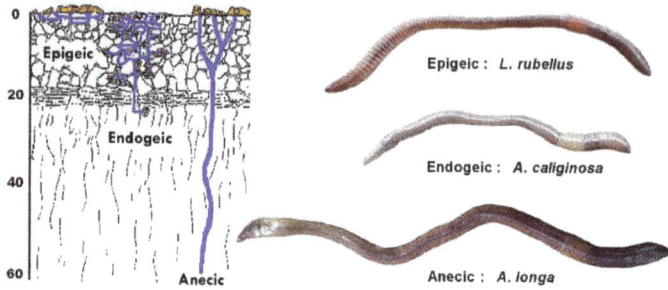

Epigeic : *L. rubellus*

Endogeic : *A. caliginosa*

Anecic : *A. longa*

1.5.1 Epigeics

1. Epigeic species grind and partially digest surface litter, rarely ingesting soil particles.

2. The soils are affected indirectly via changes in the litter, the effects of epigeic earthworms are not truly drilospheric.

3. Their mode of litter processing in natural systems results in greater nutrient leaching into the soil.

4. The epigeics feed purely on litter and generally have a short gut transit time they probably depend on a rapid response of gut microbes to aid in digestion.

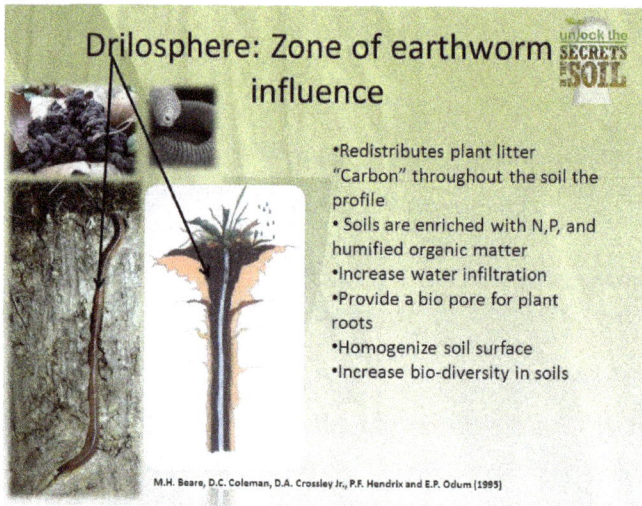

Drilosphere: Zone of earthworm influence

•Redistributes plant litter "Carbon" throughout the soil the profile
• Soils are enriched with N,P, and humified organic matter
•Increase water infiltration
•Provide a bio pore for plant roots
•Homogenize soil surface
•Increase bio-diversity in soils

M.H. Beare, D.C. Coleman, D.A. Crossley Jr., P.F. Hendrix and E.P. Odum (1995)

1.5.2 Endogeics

◉ Endogeics are the most prevalent earthworms (in biomass) in most tropical environments, often being the only group present, particularly in agro ecosystems.

◉ Endogeics are geophagous earthworms that feed on subsurface soil horizons and on soil organic matter of different qualities.

◉ They produce surface and below-ground casts.

·Live in the top 12
inches of soil
·Create extensive,
temporary, horizontal
channels to move
through in upper
layers of soil
·Aerate and mix soil
(like a natural
rototiller)
·Feed on mineral soil

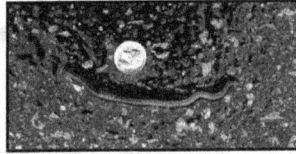

⊙ Endogeic casts with generally more clay and more organic matter than undigested soil, contain and release significant amounts of nutrients.

⊙ Fresh casts of *Pontoscolexcorethrurus*may have 2-8 times more inorganic P and NH4 than undigested soil.

⊙ This N may result from selective ingestion of richer soil portions, microbial mineralization, enteronephridial N excretions or a symbiotic N2 (Dinitrogen) fixation in the gut.

⊙ Fungal hyphae, active protozoa, algae, myxomycetes and nematodes may be digested, while encysted or protected forms survive gut passage and then rapidly proliferate in casts.

1.5.3 Anecics

⊙ Anecics the dominant earthworms (in biomass) in many temperate region soils are primarily vertically burrowing species.

⊙ They feed on surface litter and more or less are permanent refuges in underlying soil horizons.

⊙ They often produce characteristic surface features called "middens" which are circular "mound-shaped" region around a surface of the burrow's opening which is a mixture of surface organic materials and soil.

⊙ These are thought to act as "external rumens," where microbes and fauna attracted to this 'hot-spot' enhance decomposition of undigested litter and organic fragments in casts, probably due to fungal colonization of these substrates.

⊙ The feeding and casting habits of anecics may deeply influence soil characteristics up to 1m depth.

⊙ The translocation of litter, mucus excretions, air penetration and selection of soil particles enrich the burrow walls with organic matter and plant available nutrients (N, P, K and Ca).

1.6 ECOLOGICAL ROLE AND ECONOMIC IMPORTANCE OF EARTHWORMS

- The microorganisms and earthworms act symbiotically to accelerate and enhance the decomposition of organic matter.

- Earthworms in general are greatly resistant to many pesticides and concentrate the pesticides and heavy metals in their tissues.

- They also inhibit the soil borne pathogens and work as a detoxifying agent for polluted soil.

- Earthworms can be used for effluent treatment and heavy metal and pesticides removal from industrial and agricultural wastes.

- Earthworm promotes the growth of 'beneficial decomposer bacteria' in waste biomass.

- Earthworms are the most important soil invertebrate in the soil ecosystem in terms of biomass and activity, often considered as ecosystem engineer.

- Earthworms are not essential to have in the soil, but their presence can be an indicator of good soil quality.

- Earthworms improve the physical structure of soil, improve water filtration rates and absorption rates, helping the soil to drain better.

- The tunneling activity of earthworms also improves soil aeration, porosity and permeability.

- The feeding and casting habits of anecics may deeply influence soil characteristics up to >1m depth.

- The translocation of litter, mucus excretions, air penetration and selection of soil particles enrich the burrow walls with organic matter and plant available nutrients (N, P, K and Ca).

1.7 VERMICULTURE: DEFINITION, SCOPE AND IMPORTANCE

1.7.1 Vermiculture

- Vermiculture means scientific method of breeding and raising earthworms in controlled conditions.

1.7.2 Vermitechnology

- Vermitechnology is the combination of vermiculture and vermicomposting.

- Earthworm can be used for development of arable soils, break down of plant organic matter, aeration and drainage.

- Also for production of useful products like vermifertilizer and worm tissue for animal feed.

1.7.3 Environmental Requirements and Culture Methods

1. Vermicomposting – applications, future perspectives:
2. 'Gold from garbage' and 'Queen of compost'
3. Vermicomposting is a method of making compost, with the use of earthworms, which generally live in soil, eat biomass and excrete it in digested form. This compost is generally called Vermicompost or Wormicompost.

◉ **Epigeics**(surface feeders) are important in vermicomposting. The epigeics such as *Eisenia foetida*and *Eudriluseugeniae*are exotic worms and *Perionyx excavatus*is a native one being used for vermicomposting in India.

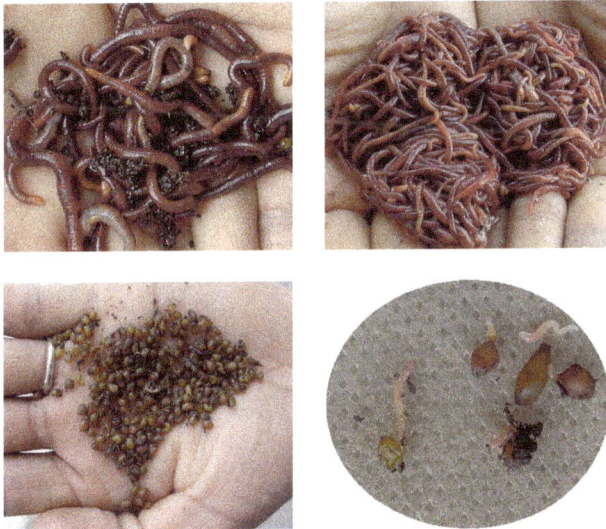

Epianecicare feeders on leaf litter and soil at upper layers of soil. This group such as *Lampitomauritii*is indigenous and is active in in-situ decomposition of organic wastes and residues in soil.

◉ Both **epigeics** and **epianecics** groups of earthworms are slender, shorter in length and red to dark brown in colour. They have high reproduction activity and efficient in recycling of organic materials.

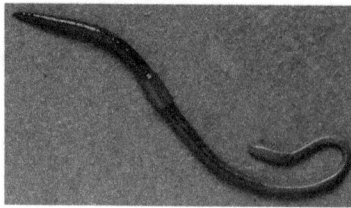

- Increased attention has been paid to *Eisenia foetida* and *Eudriluseugeniae* which have been found to be potential agent in vermicomposting of wide range of agricultural wastes and can grow at a wide range of temperature varying from 0-40 °C. However, the optimum temperature ranges from 20-30 °C.

- Materials consumed by worms undergo physical breakdown in the gizzard resulting in particles <2 µ, giving thereby an enhanced surface area for microbial processing.

- This finally ground material is exposed to various enzymes such as protease, lipase, amylase, cellulase and chitinase secreted into lumen by the gut wall and associated microbes.

- These enzymes breakdown complex biomolecules into simple compounds.

- Only 5-10% of the ingested material is absorbed into the tissues of worms for their growth and rest is excreted as cast.

- Mucus secretions of gut wall add to the structural stability of vermicompost.

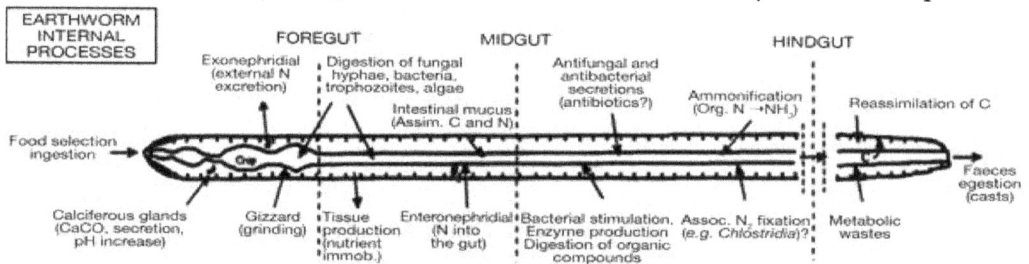

Fig.1: Diagrammatic illustration of different internal components of the drilosphere from ingestion tot excretion in earthworms – Illustration des composantes internes de la drilosphers de la digestion a l excretion chez les vers de terre.

1.8 VERMICOMPOST PREPARATION

1. Basic raw materials: Any organic material generated in the farm like wheat / rice straw, leaf fall, Paddy husk, etc.
2. Starter: Cow dung , Biogas slurry, or urine of cattle.
3. Earthworm sp.: Earth worms (Species: *Eisenia foetida*)
4. Thatched roof/vermi shed.

1.9 MICROBIOLOGY

- Microorganisms live in the alimentary canal of earthworms in a complex, mutually beneficial inter-relationship. Recently some of these microbes were isolated and identified in the gut of Eudriluseugeniae by Prabha,M.L.,et al (2014). The various types of bacteria isolated and identified were Proteus mirabilis, Staphylococcus aureus, E.coli and Klebsiella sp. and the fungi identified were Aspergillus flavus, A. niger, A. ternus, Alternaria sp. and Pencillium sp. Although these microorganisms are the same as those in the soils, the microbial population in earthworm casts is found to be much higher as compared with the surrounding soil. Earthworm casts usually have a greater population of fungi, actinomycetes and bacteria and higher enzyme activity than the surrounding soil. Microbial activity in earthworm casts may have an important effect on soil crumb structure by increasing the stability of the worm-cast-soil relative to that of the surrounding soil. Earthworms are very important in inoculating soils with microorganisms. Many microorganisms in the soil are in a dormant stage with low metabolic activity, awaiting suitable conditions like the earthworm gut or mucus to become active. Earthworms have been shown to increase the overall microbial respiration in soil, thereby enhancing microbial degradation of organic matter.

1.9.1 Microbial Action

- Earthworms make vermicompost by feeding on the waste. The other organisms which accompany them also assist in the complex process of breaking down the matter. The overall mechanism behind this is given below:

- 1. The organic matter, fungi, protozoa, algae, nematodes and bacteria ingested by earthworm is passed through its digestive tract. The majority of the bacteria and organic matter pass through undigested as 'casting' with metabolite wastes such as ammonium, urea and proteins.

- 2. During this, the worm also secretes mucus, containing polysaccharides, proteins and other nitrogenous compounds. Through feeding and excreting, worm creates a number of "burrows" in the material which helps in aeration.

- 3. Some bacteria require oxygen (aerobic) whereas some prefer its absence (anaerobic). Anaerobic bacteria are responsible for the stench from stagnant drains, landfill sites, etc. With the aerobic conditions in vermicompost, aerobic microbial growth increases. Accompanying this microbial growth is the breakdown of organic nitrogen compounds to ammonia and ammonium. The sweet smelling aerobic process overcomes the ugly smell of anaerobes. That is why worm compost piles (properly maintained) smell so nice.

⊙ 4. The whole process consumes organic matter and creates a ruffled surface in the burrow walls resulting in favourable environment for obligate aerobes (such as Pseudomonas spp., Zoogloea spp., Micrococcus spp., etc). The continued growth of the microbiological population continues to increase the rate of decomposition of the material.

⊙ 5. Air flows through the material minimize the formation of sulfide and ammonia gases, odours that are typically present in anaerobic conditions. Objectionable odours disappear quickly, due to microorganisms associated with the vermicast.

1.10. REPRODUCTION

⊙ Mature worms have a prominent band around their body, which is called as the clitellum. This is usually visible around 8-12 weeks of age. During copulation, the worms will join together at the clitellum (sometimes for quite a long period of time(Fig.4). Reproductive material is exchanged. When the worms separate, a ring of mucus material forms at the clitellum of each worm. This process is known as copulation. Sperm from the other worm is stored in sacs. As the mucus slides over the worm, it encases the sperm and eggs inside. After slipping free fromthe worm, both ends seal, forming a lemon-shape cocoon approximately 3.2 mm long (Fig.5).Two or more baby worms will hatch from one end of the cocoon in approximately 3 weeks. Baby worms are whitish to almost transparent and are 12 to 25 mm long. Red worms take 4 to 6 weeks to become sexually mature.

Fig. 4: Earthworms:Copulation

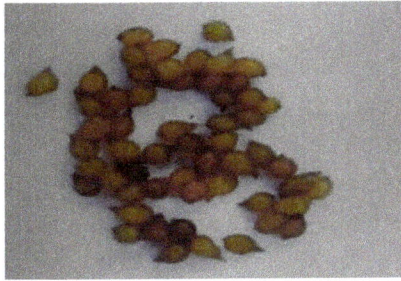

Fig.5: Earthworms: Cocoons

Calculating Rates:

Epigeic worms reproduce very rapidly and the worm populations double every 60 to 90 days, if the following conditions are provided:

- Adequate food (continuous supply of nutritious food);

- Well aerated bedding with moisture content between 70 and 90%;

- Temperatures maintained between 15 and 30ºC;

- Initial stocking densities greater than 2.5 kg/m² but not more than 5 kg/m²

The term 'stocking density' refers to the weight (initial) of worm biomass per unit area of bedding. For instance, when we start with 5 kg of worms in a bin with a surface area of 2 m², the initial stocking density would be 2.5 kg/m². The onset of rapid reproduction will be delayed at very low densities, and may even stop it completely. It is clear that worms need a certain density in order to get a chance of running into each other and reproducing frequently. At lower densities, they just don't find each other as often as the typical worm grower would like. On the other hand, densities higher than 5 kg/m2 begin to slow their productive urge, as competition for food and space increase. While it is possible to get worm densities up to as much as 20kg/m² the most common densities for vermicomposting are between 5 and 10 kg/m². Worm growers tend to stock at 5 kg/m² and "split the beds" when the density has doubled, assuming that the optimum densities for reproduction have by that point been surpassed. If the above guidelines are followed, a grower can expect a doubling in worm biomass about every 60 days. Theoretically, this means that an initial stock of 10 kg of worms can become 640 kg after one year and about 40 tonnes after two years.

The main barriers to achieving optimum rates of reproduction are:

- *Lack of knowledge and experience:* Growing worms is part science, part "green thumb". You need the knowledge, but you also need to do it to learn how to do it well.

- *Lack of dedicated resources:* Increasing worm populations requires paying attention to what is happening and responding accordingly. This takes time and effort. If the beds or windrows are neglected, the worms will likely survive, but the population will not increase at an optimum rate.

1.10.1 Life Cycle and Population

⊙ Earthworms are hermaphrodite, which means each worm is both male and female. However, each worm must still mate with another worm of its species in order to reproduce. During mating, any two adult worms can join together to fertilize each other's eggs. Fertilized egg contains in a mucous tube secreted by the clitellum that slips over its head and then into the soil through its mouth as an egg-case or cocoon. These cocoons are about the size of a match stick head and change color as the baby worms develop, starting out as pale yellow and when the hatchlings are ready to emerge, cocoons are reddish brown.

⊙ Number of cocoons and hatchling period varies for each species and depend upon the environmental conditions (Table.4). The lifespan of the earthworm in the wild is not certain, but researchers estimate a normal lifespan of about 3 years. The earthworm population is self-controlled and limited by available food, space, and environmental conditions.

Table.4: Comparison of Lifecycle and Growth of Different Earthworm Species

Earthworm	Biomass	Average reproduction rate, worm/wk	Egg Maturation period, d	Cocoon hatching d	Period to attain maturity, d	Mean matul weigl g
Eisenia fetida	0.68	10.4	85-49	32-73	53-76	0.55
Eudrilus eugeniae	5.76	6.7	43-12	13-27	32-95	4.3
Perionyx	6.3	29.4	44-71	16-21	28-56	1.3
Dendrobaena	0.16	1.4	97-214	40-126	57-86	0.92

(Source: Edwards, 1988)

1.10.2 Multiplication

⊙ Earthworms can be multiplied in 1:1 mixture of cow dung and decaying leaves kept in a cement tank or wooden box or plastic bucket with proper drainage facilities. The nucleus culture of the worms needs to be introduced into the above mixture at the rate of 50 worms per 10 kg of organic wastes properly mulched with dried grass or straw in a wet gunny bag. The unit should be kept in shade. Sufficient moisture level should be maintained by occasional sprinkling of water. Within 1-2 months, the worms multiply 300 times, which can be used for large-scale vermicomposting.

⊙ Suitability of municipal biosolids and cattle manure as substrates for vermicomposting was evaluated and reported that larger weights of newly hatched earthworms were obtained in substrate containing dry olive cake. In another study, maize straw was found to be the most suitable feed material compared to soybean (Glycine max) straw, wheat straw, chickpea (Cicer arietinum) straw and city refuse for the tropical epigeic earthworm, Perionyx excava.

1.10.3 Favourable Conditions of Earth Worms in The Composting Material

- pH: Range between 6.5 and 7.5
- Moisture: 60-70% of the moisture below and above range mortality of worms take place
- Aeration: 50% aeration from the total pore space
- Temperature: Range between 18 °C to 35 °C

1.10.4 Procedure

- It is mostly prepared in either pit or heap method.
- The dimensions either heap or pit are 10 x 4 x 2 feet.
- The length and width can be increased or decreased depending on the availability of material but not the depth because the earthworms' activity is confined to 2 feet depth only.
- First of all select a site which is not under any economic use and is shady and there is no water stagnation. The site should be near to a water source.
- 1st layer: bedding material of 1" thick with soft leaves
- 2nd layer: 9" thick organic residue layer finely chaffed material
- 3rd layer: Dung + water equal mixture of 2" layer.
- Continue the layer up to pile to ground level in the case of pit method and up to 2' in heap or surface bed method.
- Protect the worms against natural enemies like ants, lizards, snakes, frogs, toads etc., Maintain proper moisture and temperature by turnings and subsequent staking.
- At the day of 24th, 4000 worms are introduced in to the pit [1m2 =2000 worms] without disturbing the pit by regular watering the entire raw material will be turned into the vermicompost in the form of worm excreta.
- The turnover of the compost is 75% [the total material accommodated in the pit is 1000 kg; the out turn will be 750 kg]
- In-situ vermicomposting can be done by direct field application of vermicompost at 5 t/ha followed by application of cow dung (2.5 cm thick layer) and then a layer of available farm waste about 15 cm thick. Irrigation should be done at an interval of 15 days.

1.10.5 Precautions

◉ Do not cover vermicompost beds with plastic sheets because it may trap heat and gases.

◉ Do not overload the vermicompost heap to avoid high temperature that adversely affect their population.

◉ Dry conditions kill the worms and waterlogging drive them away. Watering should be done daily in summer and every third day in rainy and winter season.

◉ Addition of higher quantities of acid rich substances such as tomatoes and citrus wastes should be avoided.

◉ Make a drainage channel around the heap to avoid stagnation of water particularly in high rainfall areas in rainy season.

◉ Organic materials used for composting should be free from non-degradable materials such as stones, glass pieces, plastics, ceramic tubes/bulbs etc.

◉ The important natural enemies of vermiculture are ants, termites, centipedes, rats, pigs, birds etc.

◉ Preventive measures include treating of the site with 4% neem based insecticide before filling the heap.

1.10.6 Transportation of Live Worms

◉ Live earthworms can be packed with moist feed substrate in a container (card board/plastic) with provision of aeration.

◉ Feed substrate quantity should be roughly 0.5-1.5 g/individual for 24 hours of transportation journey.

⦿ Culture should contain cocoon, juveniles and adults because sometimes adults do not acclimatize to new environment and may even die. Under such circumstances cocoons are helpful for population build-up of earthworms.

1.10.7 Application Rate

⦿ It can be applied in any crop at any stage, but it would be more beneficial if mixed in soil after broadcasting. The rate of application is as below:

⦿ Field crops 5-6 t/ha;

⦿ Vegetables 10-12 t/ha;

⦿ Flower plants 100-200 g/sq ft;

⦿ Fruit trees 5-10 kg/tree.

1.11 ADVANTAGES OF VERMICOMPOST

⦿ Vermicompost is a rich source of nutrients, vitamins, enzymes, antibiotics and growth hormones.

⦿ Nutrient content of vermicompost is higher than traditional composts.

⦿ Vermicompost horbours certain microbial populations that help in N fixation and P solubilization.

⦿ Its application enhances nodulation in legumes and symbiotic mycorrhizal associations with the roots.

⦿ It can be used as rooting medium and for establishment of saplings in nurseries.

⦿ It has enzymes like protease, lipase, amylase, cellulase and chitinese which keep on their function of biodegradation of agricultural residues in the soil so that further microbial attack is speeded up.

⦿ It does not have foul odour as is associated with manures and decaying organic wastes.

1.12 HARVESTING OF THE VERMICOMPOST FROM THE PIT:

⦿ Stop watering before one week of harvest.

⦿ Sometimes the worms spread across the pit come in close and penetrate each other in the form of ball in 2 or 3 locations.

⦿ Heap the compost by removing the balls and place them in a bucket. However, under most instances, top layer has to be disturbed manually.

⦿ Earthworms move downward and compost is separated. After collection of compost from top layers, feed material is again replenished and composting process is rescheduled.

- The material is sieved in 2 mm sieve, the material passed through the sieve is called as vermicompost which is stored in bags.

1.11 POTENTIALS AND CONSTRAINTS FOR VERMICULTURE IN INDIA

- The warm and moist climatic conditions of India are favorable for earthworm rapid biodegradation action.

- India is an agriculture country and a large mass of people rely on agriculture, thus vermicompost is a boost.

- Enormous quantity of agricultural, natural, industrial and household wastes can be converted as vermicompost.

- The country will become clean and green.

2

COMPOST: THE BLACK GOLD

2.1 INTRODUCTION

Soil, the uppermost crust is the earth's fragile skin that anchors all life on Earth. The fragile skin that supports the life on this planet is significantly affected by various forms of soil degradation. Half of the topsoil on the planet earth has been lost in the last 150 years. Soil degradation is estimated to be occurring on 147 million hectares (Mha) of land, including 94 Mha from water erosion, 16 Mha from acidification, 14 Mha from flooding, 9 Mha from wind erosion, 6 Mha from salinity, and 7 Mha from a combination of factors in India alone. It is very severe because India supports 18% of the world's human population and 15% livestock population, but has only 2.4% of the world's land area. Despite its low proportional land area, India ranks second worldwide in farm output Bhattacharyya, 2015). Land degradation and poor soil quality are the major threats to our food and environmental security. Hilly terrains and high rainfall regions are more vulnerable to soil degradation. Soil erosion in these regions causes loss of top fertile soil, organic matter, nutrients and other beneficial microbes and adversely affects soil health. Organic matter helps in reducing soil vulnerability and protects nutrient loss apart from nutritional supplementation.

The practice of gathering and using wastes from animal, human and vegetable sources for improving crop productivity is as old as agriculture. These organic wastes offer the possibility of sustaining crop yields and maintenance of soil health. It also helps in evading the use of chemical fertilisers; pesticides etc. as it supplies the required nutrients. During the growth and production, crops uptake the nutrients from the soil, and the same has to be replenished to sustain the soil productivity. The best way of replenishing is to supply the nutrients received from decomposed and digested organic residues available. The process of decomposing plant residues and other non-living organic materials to make an earthy, dark, crumbly substance, which is excellent to enrich and replenish the soil is called composting and the product thus obtained is known as compost or humus. Compost is a way of returning safe, easily mineralised, organic matter, or humus, to the soil. Though the organic matter is a relatively small fraction of soil, it can have a dynamic influence on the health of the soil. Since the compost is black and can enrich the soil and it also supplies the nutrients required by the crops to enhance its performance so, it can be termed as "The Black Gold".

Fig. 1: Finished Compost 'the black gold'

Compost can provide valuable nutrients and organic matter to the soil, depending upon the composting approaches and raw materials (feedstock) used. Chemical analysis of different kind of compost samples indicates its total nitrogen status, available nitrogen, phosphorus, and potassium quantities, but most of them are relatively low in one or more nutrients and are not considered good "fertilisers"; however, as soil amendments, they are good sources of organic matter.

Nitrogen and phosphorous in compost are generally found in both plant-available forms (NO_3, NH_4, and P_2O_5) and organic forms. The nutrient present in organic forms will be converted to "plant- available" form after decomposition. Therefore, readily available nutrients in compost can be much lower than in raw waste, but a "timed-release" effect occurs in compost. This slow-release of nutrients "bound" initially in organic forms makes compost nutrient efficient because it prevents the loss of nutrients, which generally occurs in fertilisers.

2.2 PRINCIPAL ADVANTAGE COMPOSTING

The principal advantage of making compost is utilising the nutrients available in the waste as a supplement to crop production. Primarily composting stabilises organic materials so that they can be stored safely, transported easily, and applied at a convenient time. Compost can also be converted into marketable product and may be sold in urban and per urban areas.

Large-scale composting is practised by municipal wastewater treatment plants, agricultural producers, industrial waste generators, and commercial composters (who are in the business of composting wastes collected from various sources). These bulk producers either sell their product to nurseries, big farms, landscaping companies, etc. or may be filled in 1-5 kg packets for retail selling to the general public. Depending upon compost production and markets, prices of commercially available compost may vary considerably.

Raw organic waste may be applied directly to farmland, but there are limitations of application timings. Further maintaining optimum moisture and temperature for proper decomposition on entire farmland may not be feasible; thus, there will be improper decomposition and nutrient loss. Apart from this scattering, the wastes to the large area may cause unhygienic environment, lousy odour and pollution. On other hands, compost is easier to handle and store than raw waste. It does not have any offensive odour and is less likely to contribute to water contamination. Simultaneously composting of organic waste also provides flexibility to the farmer and he can choose the time and mode of application. Only problem farmer may have that is storage, as compost being bulky requires vast space for storing. So, the information given below may be helpful to the farmers to decide whether composting will be appropriate for them.

2.3 PREREQUISITES OF COMPOSTING

The feasibility of composting depends upon the availability of raw materials (feedstock), composting space, composting equipment and labour. Composting is a biological process where microorganisms (bacteria, fungi, and actinomycetes) breaks the complex organic materials into simple and more stable forms. They feed on organic waste during their growth and development, and during the process, they decompose the organic matter into simpler compounds that can be utilised by them. Just can be said that we are feeding a balanced diet to these organisms and providing them with a comfortable environment to make them work efficiently.

So the availability of good quality feedstock having a balanced amount of nutrients for composting is of prime importance. Excreta and wastes from livestock farms are the rich sources of macro and micronutrients that can be composted as it is, but the mixing of low nutrient feedstock like straws and other crop residues can make the feedstock a balanced foodstuff for microorganisms. Moreover, livestock excreta is very rich in nitrogen, and if composted alone, the nitrogen present may convert into ammonia, which may be lost through volatilisation. Sometimes very strong and pungent odour one may feel near compost pit, this may occur due to volatilisation of ammonia. Mixing of high carbon – low nitrogen feedstock with livestock farm waste balances the carbon: nitrogen (C:N) ratio. A C:N ratio between 20:1 to 30:1 is a good ratio for proper microbial activity and making good quality compost. Compaction of feedstock during the composting restricts aerobic breakdown and creates anaerobic conditions. It may lead to the production of offensive smell and harmful gasses like methane etc. Crop residues, straws, sawmill dust, wood chips, leaves etc. can be added to increase aeration and reduce the bulk density of the composted mass.

Composting area is another essential aspect to look upon. As the compost and the feedstock are too bulky, so transportation becomes an important factor. The composting area should be near to the place where plenty of wastes and feedstuff is available, as it reduces the cost of transportation. However, the composting pit should be located sufficiently away from the residential area as the sometimes unpleasant smell is produced due to faulty decomposition. Lousy odour and flies from feedstock and compost piles may be offensive to residents of the area.

Home scale composting does not require much equipment but, it can be done with small household equipment like, spade, shovel, rakes etc. However, for commercial scale, machinery is required for transporting, mixing and composting of the feedstock. The following equipment/machinery are essential for the commercial composting unit:

- Truck for transportation
- Front-end loader
- Shredder
- Turner
- Aerator
- Composting thermometer
- Shovel

2.3.1 Crop Residue Availability

Crop residues are the remains of plants that are left in the field after harvesting and thrashing of the crops. These residues are a good source of plant nutrients, as quite a good amount of the nutrients absorbed from the soil remains in the residues. These residues can be converted into the compost of better quality, and its use can help to sustain or even improve crop yield. Therefore, efficient crop residue management can play a vital role in restoring soil productivity as well as in increasing the fertiliser use efficiency (FUE).

Organic wastes are generated regularly at the farm as well as at household levels, and disposal of such waste is a serious problem. These wastes either burnt or used as landfills, which is a poor utilisation of a very useful resource. Many times improper disposal of waste may become hazardous, and it may contaminate the ecosystem. On other hands, these organic waste can be a good source of nutrient when decomposed properly. The nutrients present in these wastes can be effectively used as organic manure for increasing the agricultural productivity besides having environmental benefits.

2.3.2 Burning of Crop Residues in Fields

Farmers use residues for their own needs or sell to other landless households. Farmers intentionally burn the surplus residues for clearing of the fields, fertility enhancement in the form of ash addition and pest and pasture management.

Fig. 2: Burning of crop residues

Farmers also perceive that burning kills harmful pathogens. Residue burning increases the short-term availability of some nutrients, i.e., Calcium and potassium and reduces soil acidity but leads to loss of other nutrients like nitrogen and sulphur, organic matter and damages microflora in topsoil.

2.4 WAYS TO IMPROVE SOIL HEALTH

An increase in biomass production improves root biomass proportionately, a significant component of which goes back into the soil. Soil organic carbon enhancement through crop residue recycling, inclusion of legumes in the cropping sequence or as intercrops, green manure crops, green leaf manuring, tank silt addition, farmyard manure, biofertilizer, composting/ vermicomposting along with fertilizers and integrated nutrient management are some of the important options to improve soil health and crop productivity in rainfed areas.

Composting of agricultural waste is an important method in which organic waste such as food, leaves and paper is turned into a material that is useful to the environment. Micro-organisms and bacteria break down the waste to form a paste-like substance. The resulting material is rich in nutrients and oxygen. Composting is becoming an effective way to increase organic matter of the soil. In addition to increasing organic matter of the soil, amending with compost also increases soil microbial populations, which leads to an improvement in the soil quality. Composting comes in many different forms.

2.5 COMPOST QUALITY

Compost quality varies with the kind of raw organic materials (feedstock), the composting process used, and the state of biological activity. Before using compost as a soil amendment, it is a good idea to evaluate its quality by determining moisture content, organic matter content, carbon to nitrogen (C: N) ratio, and pH (Table 1).

Table 1. Qualities of compost for on-farm use and methods to test (Cooperband, 2002).

S. No.	Quality	Optimum	How to test
1.	Source of organic matter	Should have a good organic matter content (40-60%)	Have organic matter tested by a soil lab
2.	Source of nitrogen	10–15:1 C:N ratio	Have C:N ratio testesd by a soil lab
3.	Neutral pH	6–8	Use soil pH kit to test pH at home or have pH Tested by a soil laboratory
4.	No phototoxic compounds	Good seed germination (>85%)	Plant 10 seeds in a small pot
5.	Weed-free	No or few weed seeds	Moisten compost and wait for weed seedlings to grow

2.6 COMPOST AND SUPPRESSION OF PATHOGENS

Compost helps in developing a suppressing soil that favours the healthy growth of plants and suppresses the establishment of pathogens. The mechanisms which suppress disease organisms in these soils include induced resistance, direct parasitism (one organism consuming another), nutrient competition, and direct inhibition through antibiotics secreted by beneficial organisms.

Compost enhances vigour and resistance ability of the plant for any pathogenic attack. On other hands, it also favours the growth of microflora that may parasitise over the pathogens or may produce antagonistic microbial chemicals, including the production of antibiotics and parasitism, which suppresses the growth of pathogenic organisms. Organic acids and ammonia present in the compost have anantagonistic effect on pathogens. Compost also produces some compounds that stimulate premature germination of pathogens resulting in reduced pathogenic load.

2.7 COMPOST USE AND SOIL FERTILITY

Compost is a nutrient-rich soil conditioner that improves soil quality. It not only supplies nutrients but influences the availability of nutrients to the plants and improves the soil health primarily by improving physical and biological properties of soil. Compost has along-term impact on soilnutritional status, soil texture, soil structure, soil erosion and water dynamics. It reduces water runoff, soil detachment and transfer of ammonium and nitrate ions to the water. Thus the application of compost besides supplying nutrients also increase overall soil fertility.

2.8 COMPOST APPLICATION

The timing of application is an important factor to get the best result. Timing greatly depends on edaphic factors. Compost supports the microbial activity, so the best time of compost application is before the onset of the monsoon is the most favourable time for the microbes. However, in an area of high rainfall, spring is the best time to apply it, to avoid leaching of some of the valuable nutrients.

Compost releases nutrient slowly, and the nutrients may not be available for uptake by the plant immediately after application. Most of the nitrogen remaining after completion of the composting process is bound into organic forms and releases slowly. Compost application rates can be calculated using fertiliser recommendations from soil tests and compost nutrient analysis. While calculating the quantity of compost based on soil test report and crop requirement, then it must also be taken into account that only 10 to 25% of Nitrogen (N), 40% of Phosphorus (P2O5) and 60% of Potash (K2O) present in compost will be available to plant during the first year of application. It is important to know that the actual availability of nutrients will depend on the nature of the raw material used for composting and environmental conditions.

2.9 DEGENERATION AND DECAYING OF WASTES A NECESSARY EVIL

The common practices of disposing of waste are dumping it from out of sight of ordinary people. However, it does not solve the problem but indirectly increases the same manifold and sometimes it goes beyond the control of everybody. The consequences of this practice such as health hazards, pollution of soil, water, air & food, unpleasant surroundings, loss of precious resources that could be obtained from the solid waste, etc. are well known (Agarwal et al., 2015). Many times when we roam in any village, we may encounter the heaps of partially rotten organic residues left beside the pathways emitting some off smell and polluting the environment. The situation further darkens when the village does not has adequate sanitation facilities as a rising population, and declining vegetation cover is demanding. This scene does not create the positive image of the people residing and on other hands, it also shows the lack of awareness about the ways to convert the waste into a high-quality product the BLACK GOLD or say compost.

The decomposition of waste into constituent chemicals is a common source of local environmental pollution. This problem is quite acute in developing nations. The release of foul-smelling gases is a major environmental concern by decomposing garbage. Methane released from decomposing garbage is a by-product of the anaerobic respiration that contributes to the greenhouse effect and climate change. During degeneration process, the organic material is subjected to the activities of many micro- organisms. Often the disease-causing organisms find rotting garbage as a safe place to multiply, and then they spread all around causing diseases in human beings, animals and plants.

This decomposition is a necessary evil, though it looks unpleasant, it helps in cleaning. Imagine if there would not have been this decomposition and decaying process then what would have been happened. The land where we are living would have been full of residues and dead. Even the faeces would not disappear from the earth. We can easily imagine what a possible place the earth will become. No plant will be able to absorb any nutrients from the soil or synthesise any food through photosynthesis, nor can any animal or human being eat and digest any food. Therefore, decomposition or decaying is virtually a blessing in disguise, and it had made the earth a beautiful place for living and cultivating, as nature has created the process to clean the environment so the new one may have sufficient space to dwell. Since decomposition is necessary and unavoidable, so it is important to handle it properly. Therefore, decomposition of biomass/organic material may be done in the presence of oxygen using micro-organisms that will not create unsanitary conditions and even a valuable product as organic manures will be obtained.

2.10 PRINCIPLES OF COMPOSTING AND CHANGES DURING COMPOSTING

Composting is a method of aerobic decomposition of crop residues, wastes from livestock farms and other non-living organic materials and converts them to an earthy, dark, crumbly substance that can enrich and replenish the soil. In this process, various microfauna such as nematodes, mites, spiders, centipedes, earthworms, ground beetles etc. and microorganisms, including bacteria, fungi and actinomycetes break down organic matter into simpler substances in the aerated environment and give the finished product compost or humus. Efficacy of the composting is influenced by the environmental factors such as aeration, temperature, moisture, substrate, microbial populations etc.

Growth and reproduction is a feature of all the livings, so the micro- organisms present there also grow in size and increase in number and during this process they require nutrients, oxygen and moisture, which they obtain by decomposing the organic residues and surroundings. They get the large amount of energy, carbon, nitrogen, phosphorus, macro and micro minerals from the residue itself, whereas oxygen and moisture, they obtain from the surrounding atmosphere. Complex organic compounds of the residue are broken down into carbon dioxide, water vapour and energy. Some portion of this energy is utilised by the micro-organisms to carry out their life processes, but a significant portion of it is converted into heat, which increases the temperature inside the compost pit or heap. The temperature of the pit or heap goes as high as 60-700C that is sufficient enough to destroy many diseases causing agents, harmful insects and weed seeds. Carbon dioxide which is produced in a large amount leaves the pit along with some vapour, and consequently, the volume reduces. Temperature reduces again when the decomposition is over. Dark and crumbly substance remained after decomposition of the organic matter has humus, other broken-down products, and living and dead microbial cells called compost and is ready for use.

2.11 ESSENTIAL CONDITIONS THAT ARE THE PREREQUISITES FOR OBTAINING QUALITY COMPOST

2.11.1 The Type of Organic Matter

Generally, everything of plant or animal origin can be used for the composting but for achieving better results; it is important to know the combination of materials that can be used. Composting microbes use carbon as energy source and nitrogen for growth (protein synthesis). When the substrate is filled in the compost pit, it is essential to maintain a proper balance of carbon to nitrogen (C:N ratio). The ratio of the available carbon (C) and nitrogen (N) plays an important role in decomposition, the organic matter with higher nitrogen content decomposes faster than the low nitrogen containing material. The proportion may vary; at C:N ratios from 25:1 to 40:1 for proper decomposition. A mixture of materials containing 30 parts of carbon to 1 part of nitrogen is considered ideal for compost.

Based on the nitrogen content, the organic matter may be grouped as low nitrogen containing or "high C:N ratio" and high nitrogen containing or "low C:N ratio" matters. The first group consists of rough materials and decomposes slowly, whereas the second group consist of dung, animal wastes and young juicy plant parts, and decomposes readily. The list of the organic residues grouped in these two broad categories is given in Table 1.

Table 1: C:N ratio or different composting materials

High C:N Ratio (> Nitrogen)	Low C:N Ratio (< Nitrogen)
Straws of paddy, wheat, oat, rye and millets etc., rice hulls	Animal Manures
Sugarcane leaves trash and baggage	Dried blood, Slaughter house wastes, Hoof and horn meal, Bonemeal
Pigeon pea stalks	Sour milk, Urine
Old green leaves, dried leaves, non-legume hay and twigs of plants	Fish cleaning/meals, chicken feathers
Mature grasses	Brewers waste
Coconut fibre waste	Kitchen scraps
Peanut hulls, Potato waste	Water hyacinth, seaweeds,
Mustard plants (after harvest)	Fresh young grass
Paper, newspaper, cardboard	Young plants
Sawdust, tree bark, wood chip	Legume Hay
Silk mill wastes	Tea/Coffee Grounds
Egg shells, ashes	Cotton, Soybean, Corn Meal

Higher nitrogen content facilitates rapid multiplication of microorganisms thus substrate decomposes at a faster rate, whereas substrate with low nitrogen decomposes slowly as multiplication of microbes remains slowly.

Table 2: Effect of C:N ratio the pace of decomposition

Pace of decomposition	C:N Ratio
Fast	• C:N ratio between 20/1 and 40/1 • Heats up quickly • Generally, two turnings total • Ready in 2 to 3 months

Medium	• C:N ratio closer to 100/1 takes longer to mature (about three months in warm weather, nine months in cold weather).
	• Turning is optional but beneficial
Slow	• C:N ratio around 200/1 or higher
	• It generally needs to be mixed later with
	• another pile

2.11.2 Organisms

Numerous organisms take part in the decomposition of organic residues. Depending on their size, they may be termed as macro- organisms like; mites, centipedes, snails, millipedes, springtails, spiders, slugs, beetles, ants, flies, nematodes, flatworms, rotifers and earthworms or micro-organisms such as bacteria, fungi, and actinomycetes. The macro-organisms are considered to be physical decomposers because they grind, bite, suck, tear, and chew materials into smaller pieces; however, micro-organisms are considered chemical decomposers, because they change the chemistry of organic wastes and they account for most of the decomposition that takes place in the substrate.

Aerobic bacteria are the most important decomposers among all the organisms. They are abundantly available and may count millions in a gram of soil or decaying organic matter. These bacteria are the most nutritionally diverse of all organisms and can eat nearly anything. They utilise carbon as a source of energy and nitrogen to build their protein for multiplication. They oxidise carbon fraction of the organic substrate to obtain energy. During oxidation temperature of the compost pile increases from ambient atmospheric temperature. The rate of temperature increment depends upon the composition of decomposable materials. Micro-organism can survive in an extended range of temperature from 0 to 80 °C but, majority of them are active in decomposition within the temperature range of 30 to 40 °C. While bacteria can survive in unfavourable environments as they can't escape but changes in pH or fluctuation in environmental factors like; oxygen, moisture, temperature etc. can make bacteria die or they become inactive. Aerobic bacteria, which are most preferred microorganism for rapid decomposition of organic matter, need oxygen levels higher than five per cent for better decomposition. When oxygen levels fall below five per cent, the population of the aerobes reduces and decomposition process slows by as much as 90 per cent. In this condition, anaerobic micro-organisms take over the process, and they start producing a lot of useless organic acids and amines that makes many nutrients including nitrogen unavailable to the plants. These products include; hydrogen sulfide (smell like rotten eggs), cadaverine, and putrescine (offensive odours) produce the rotten and stinky smell and in some cases, are toxic to plants.

Diverse kinds of aerobic bacteria act on the substrate, and their respective population varies with the temperature of the pile. Psychrophilic bacteria work at the very low-temperature range and may remain active at temperature lesser than 20 ^{0}C. Though they produce quite a small amount of heat that is enough to enhance the temperature of the pile. As temperature increases more than 20 0C, mesophilic bacteria, start to take over. Mesophilic bacteria rapidly decompose organic matter and produces acids, carbon dioxide and heat. Their working temperature range is generally in between 20 to 37 ^{0}C. When the temperature of the pile increases further, the mesophilic bacteria begin to die off or they move to the outer part of the heap. At about than 40 ^{0}C thermophilic bacteria take over as they thrive well at temperatures ranging from 45 to 70 ^{0}C. Thermophiles continue the decomposition process, till the pile temperature reaches 65 to 70 ^{0}C, where it usually stabilises. Feeding of new material and regular turning sustains high temperature or else this temperature lasts for three to five days. The high temperatures (above 60 ^{0}C) have the advantage of killing pathogenic organisms and weed seeds. If the temperature exceeds than 70 ^{0}C, then action should be taken for cooling the heap by turning it, because the temperature of the pile beyond 70 ^{0}C, makes the composting material sterile and it loses its nutritional qualities and disease-fighting properties.

As the temperature reaches beyond 70 ^{0}C, the population of thermophilic bacteria starts declining, and that results in the gradual reduction of the temperature of the pile. As the pile cools off, the mesophilic bacteria again become dominant, and they start consuming remaining organic material with the help of other organisms and now composting process enters in another phase where actinomycetes and fungi play their role.

While the numerous types of bacteria are at work, other micro- organisms are also contributing to the degradation process. Greyishappearing actinomycetes, a higher-form bacteria are similar to fungiand moulds. They are responsible for the pleasant earthy smell of compost. Actinomycetes decompose more resistant materials in a pile such as lignin, cellulose, starches and proteins, and reduce them to carbon, nitrogen, and ammonia, making nutrients available for higher plants. Actinomycetes occur in large clusters and become very evident during the later stages of decomposition.

Like bacteria and actinomycetes, fungi are also contributing to the decay of organic matter in a compost pile. Fungi are primitive plantsthat may be either single or multicellular and filamentous. Theylacka photosynthetic pigment and are responsible for the breakdown cellulose and lignin. They prefer cooler temperatures (22 to 25^{0}C) and easily digested food sources, so they take over the process during the final stage of composting.

Table 3: Organisms in composting

S. No.	Group	Organisms	Action
1	Macro-fauna (small soil animals)	Mites, ants, termites, millipedes, centipedes, spiders, beetles and worms.	Grind, bite, suck, tear, cut and chew materialsinto smaller pieces.

S. No.	Group	Organisms	Action
2	Micro- fauna (small multicellular organisms)	Rotifers	Rotifers are microscopic organisms found in films of water in the compost, and they help in controlling populations of bacteria and small protozoan
3	Micro-fauna	Microscopic animals like Protozoa	Help in consuming bacteria, fungi and micro-organic particulates.
4	Micro-flora (Single-celled)	Bacteria,	The common of all the microorganisms found in compost
5	M i c r o - f l o r a (Multicellular and filamentous)	Actinomycetes,	Necessary for breaking down paper products such as newspaper, bark, etc.
6	Macro-flora (larger plants)	Fungi, moulds, yeasts, algae, viruses	Breaking down of materials that bacteria cannot, especially lignin in woody material.
7	Macro-fauna (Larger soil organisms)	Earthworms	Ingestion of partly composted material, but also continually re-create aeration and drainage tunnels as they move through the compost.

2.11.3 Aeration

Composting process accomplishes in the presence of macro and micro-organisms, and adequate supply of oxygen is essential for them to carry out respiration to sustain their life process. Therefore, the heap should be prepared in such a way that it provides adequate aeration. Anaerobic conditions (absence of air) leads to the development of different unfavourable kinds of micro-organisms, causing putrefaction of residue and degrading the quality of the compost.

2.11.4 Moisture

Every living being requires water to sustain their life and the organisms engaged in composting, too need sufficient moisture to sustain and grow. Microbial activity occurs most rapidly in thin water films on the surface of organic materials. Micro-organisms can only utilise organic molecules that are dissolved in water. The optimum moisture content for compost pile ranges from 40 to 60 per cent. If moisture reduces from 40 per cent, then bacterial activity reduces, and they may become dormant, however, if moisture increases from 60 per cent then it forces air out of pile pore spaces, suffocating the aerobic bacteria. The population of aerobic bacteria reduces and decomposition is taken over by anaerobic bacteria, resulting in putrefaction and unpleasant odours.

Special care is required to maintain adequate moisture in the composting material in tropical regions, whereas in temperate regions the chances of water losses are scanty. Wetting of the mixture initially and at each turning, using artificial windbreaks and shading may help in reducing the water losses; however during monsoon, the heap may be built above ground at an elevated site to control the excess of moisture.

2.11.5 Temperature

Soon after putting the material in a heap, rapid decomposition takes place. The heap passes through all the stages of warming-up, high temperature, cooling down and maturing. In the beginning, basic complex organic compounds like starch, sugars and fats are broken down, and the heat generated during this process warms up the heap soon after it reaches a peak of 60 to 70 ^{0}C. At peak stage, loss of heat from the heap is more or less equal to the amount of heat generated by the micro-organisms. The peak period of heat in a heap is essential for the destruction of pathogenic organisms and weed seeds. It generally occurs after 5-8 days of heaping or pitting. The temperature in the middle of the pile goes up to 70 to 75 ^{0}C and gradually cools down. However, the optimum temperature is maintained at 600C during 10-15 days after pitting, then slowly comes down to about 200 °C.

Compost thermometer may be the best way to monitor the temperature, but one may judge the temperature of the pile by merely putting his/her fist into the pile. A metal pipe or iron bar may also be inserted in the middle of the pile, periodically pulling it out and feeling it may also give an idea about the temperature of the heap. If the bar is hot or the interior of the pile feels uncomfortably warm or hot during the first few weeks of composting, one may know that everything is fine. If the temperature inside the pile is the same as the outside that is an indication that the composting process is slow. Then adding nitrogen-rich material and turning the pile may increase the decomposition rate.

Outside air temperature also impacts the decomposition process. Warmer outside temperatures during late spring, summer and early monsoon stimulates bacteria and speeds up decomposition, whereas low winter temperature slows down or temporarily stops the composting process. As air temperature warms up in the spring, microbial activity will resume or else during winter months, compost piles can be covered with polythene or a tarp to retain heat longer.

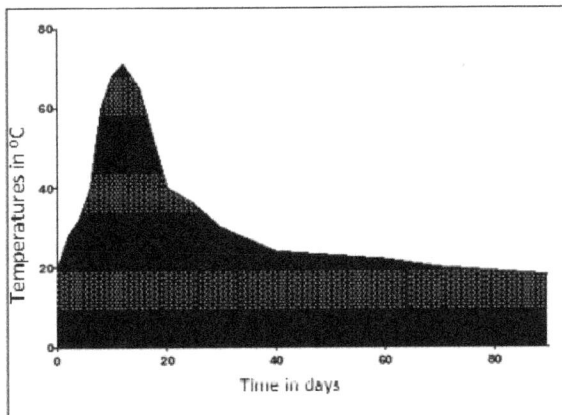

Fig. 3: Temperature curve for an unturned compost pile

2.11.6 Reaction or pH

The initial pH of compost heaps is slightly acidic, i.e. around pH 6 as is found in the cell sap of most of the plants. Compost micro-organisms also operate best under neutral to slightly acidic conditions, with a pH range of 5.5 to 8. During the early stages of decomposition, organic acids are formed, and this acidic condition favours the growth of fungi and disintegration of lignin and cellulose. As composting proceeds, the organic acids become neutralised, and mature compost generally has a pH between 6 and 8.

Compost heaps are rarely too alkaline. However, in some instances when composting is highly alkaline, loss of nitrogen through volatilisation takes place. Whereas, highly acidic conditions, in the beginning, causes reduced microbial activity and the heap fails to warm up. Adding lime (CaCO3) generally is not recommended because it causes loss of ammoniacal nitrogen to the atmosphere in the form of ammonia gas. This loss not only causes stringent odour but also depletes nitrogen that is better kept in the compost for future use by plants. Use of household ashes or egg shells helps in preventing too much acidity. Ashes must be used if young and succulent materials are predominately used for composting. Usually, if careful attention is paid in the making of the heap, especially in moistening the content and aeration, then acidity or alkalinity will not be a problem. If anaerobic conditions develop during composting, organic acids may accumulate rather than break down. Aerating or mixing the system reduces this acidity.

2.12 CHANGES DURING COMPOSTING

Anything that is alive or once it was alive, constitute the organic matter, and all the sources for making compost comes from there. The plant and animal residues are primarily made up of sugars, starch, cellulose, hemicellulose, lignins, pectins, resins, proteins, fats and waxes. When these waste materials are placed in heap or pit for composting, they are attacked by a variety of macro and micro- organisms including bacteria, fungi, actinomycetes, protozoa, worms and insects larvae etc. As a result of these activities, a considerable portion of the constituent compounds available in the residues is degraded from their original complex forms to the new simple soluble forms. These degraded materials may be solids or liquids like phosphate, potassium, ammonium, nitrate, organic acids, etc. or gases like carbon dioxide, methane, hydrogen sulphide, hydrogen and ammonia. Among the organic fractions, those which readily lend themselves to decomposition are the celluloses, hemicellulose, proteins, waxes and other nitrogenous substances. Apart from this, the dark and crumbly bulk remained after decomposition of the organic matter called humus, and it forms the major portion of organic residues. The remained organic residue, or humus contains not only the remains of the organic substrate used but also dead and decomposed part of all the macro and micro-organisms which were involved in decomposition of the matter. Finally, the finished product is called the compost that may be termed as BLACK GOLD because of its nutritional advantages in replenishing the lost soil fertility.

3

VERMICULTURE BIOTECHNOLOGY

3.1 INTRODUCTION

The word biotechnology means theoretical approaches based on knowledge through which living organisms may be utilized commercially and the word Vermiculture mean the scientific method of breeding and multiplications of earthworms in controlled conditions. It aims at creating improved conditions artificially so that the earthworms can multiply in short est period of time and space.

3.2 USE OF VERMICULTURE BIOTECHNOLOGY

Vermiculture biotechnology will be useful in field for-

- City garbage recycling deodorization and to keep environment clean.
- Management of effluents from intensively housed livestock.
- Management of sewage sludge.
- Decomposition of organic, industrial waste matters especially papermill waste.

3.3 BENEFIT OF VERMICULTURE

The major economic benefits that can be accrued from Vermiculture areas follows:

1.	Soil turnover.
2.	Improvement of soil aeration.
3.	Increased rate of humification.
4.	Agriwaste resource cyclin
5.	Once the soil sare in oculated with earthworms, it need not be repeated as the multiplication of worm goes on every year.
6.	Stimulation of respiratoryandenzymatic activity of soil.

7.	Reducing soil erosion. Worm action increase swater in filtrati on capacity of soil and thuslessrun off.
8.	Enrichment of soil fertility.
9.	Help in ginreclamation of wasteland.
10.	Increase in yield.
11.	Irrigation requirement of cropplants reduces because:
i)	Water is stored to a greater depth in the soil.
ii)	Water holding capacity of soil increases due to vermicastings, which also absorb moisture from the air particularly during night.
iii)	Mulch layer on ground reduces moisture loss from soil .
12.	Significant increase in N- fixing bacteria.
13.	Conversion of organic N and P in plant assimilable form.
14.	Destruction of phytopathogenic fungi.
15.	Assimilation of nematodes.
16.	Bactericidal and bacteriostatic properties of coelemic fluids of earthworm.
17.	Earthworm ability to eliminate offensive odor.
18.	Conservation of water. Protecting ground water from getting polluted.
19.	As protein source for poultry, fishery, pigs, pets etc. Earthworm body contains 70% crude protein , higher than fish meal (65%), meat and bone meal (50%), and soyabean meal (45%).
20.	Extraction of enzymes, medicinal compounds and high value amino acids.
21.	Production of Vermicompost which may be developed as an enterprise for the source of income.

3.4 WHAT EARTHWORMS DO?

Earthworms live in the soil and perform following role-

- Earthworms eat plant biomass,which fall on the form of mulch.
- Earthworms are effective tool for speedy development of wastelands.
- Earthworms recharge ground water.
- Earthworms maintain soil aeration.
- Earthworms produce10 mm soil per year, which is produced by naturein200 years.
- Earthworms maintain soil temperature and moisture.
- Earthworms increase root volume and bacterial activity about 10 fold.
- Earthworms feed on soil and soil organic matter and convert it to compost,making the soil rich in nutrients.
- Earthworms are natural tillers of soil.

- Earthworms aerate and pulverize soil.

- Earthworms make soil porous, improving drainage.

- Earthworms increase water-holding capacity of soil.

- Earthworms encourage growth of useful microorganisms, which also make soil rich.

- Earthworms produce enzymes, hormones,vitaminsandantibiotics,thereby increasing immunity of plants against some pests.

3.5 TAXONOMIC POSITION OF THE EARTHWORM

Earthworms are segmented animals, hermaphrodite (bi-sexual), invertebratecreatures belong to Phylum–Annelida (Bilaterally symmetrical, metamerically segmented, true coelomates), Class – Oligochaeta (Setae embedded in the integument,parapodiaabsent) and Order-Terricolaeor Neooligochaeta (Clitellumin 14th, 15th and 16th segments, and eye spots absent). Earthworms found burrowing in moist soil, sand ormud all over the world. Their body is elongated, cylindrical and pointed at both ends, It is adapted for burrowing. Body is metamerically segmented,consistingofabout120metameres (segments).Prostomium reduced, forms upper lip and hangs in front of mouth. A ring of setae present in each segment except the first and last segments. Female genitalaperture unpaired and present in 14th segment but male genital apertures are paired and present in 18th segment. Two pairs of genital papillae present on the ventral surface, one pair in each of the segments 17th and 19th. Anus situated in the last segments and they reproduced usually sexual. Earthworms have no eye but photo sensitive i.e. respond to light, have no ear extremely sensitive to touch and vibrations, have no teeth but feeds organic materials, have no nose but take oxygenthrough skin. Amatured red earthworm lay cocoons (eggs) every 7-10 days. Cocoons incubate within 14-21 days in moist bedat 20 ^{0}C–30 ^{0}C. After hatching immatures develop to reproductives tage within 60-90 days with 4-8 inch length. One earthworm produces with in a period of 12 months about100-200worms.

3.6 TYPE OF EARTHWORMS

- Charles Darwin (1881) carried out first systemic study on earthworms. Later on George Oliver and Barret taken the clue from Darwin's work and demonstrated that earthworms could be utilized to draw excellent benefits in agriculture. Barret (1947) was the first commercial worm grower and operated on a tonnage scale. Barret's success encouraged many others but it was only in the late sixties that some scientists felt the need to merge ecology with the concept of utilizing earthworms commercially to draw multifarious benefits.

- There are about 386 varieties of earthworm. Earthworms can be classified into three types viz; Epigeic, Endogeic and Anisic (Diagelc) according to their feeding habits, distribution in soil strata, defecation activities, and response to soil constraints. Worm can multiply 20 to 25 times with in 65 to 70 days in favourable conditions.

3.7 VERMICOMPOSTING

As regards the usefulness of earthworms in alternative agriculture, it is suffice to consider Vermicomposting and increasing earthworm population of all the species of worms in the fields.

Vermicomposting is the bioconversion of organic waste materials into nutritious compost by earthworm activity. Vermicomposting is an easy and effective way to recycle agricultural waste, city garbage and kitchen waste. In this process worms help in transforming waste into high quality fertilizer. There are reports that, in general, Vermicompost helps to boost productivity of crops by 40% even at 20 to 60% lower nutrient input. The Vermicast is valuable soil amendment and may replace the chemical fertilizer. Therefore, many countries have adopted Vermicomposting for sound waste management and recycling strategies. In Vermicomposting process, earthworms are used to produce Vermicompost from variety of organic waste mixes. In this process the technology employed is very simple and can be easily handled even by an uneducated, unskilled farmwoman after a brief training.

3.8 VERMICAST

After passing through the earthworm gut, ingested soil is expelled as globular soil aggregates called Vermicast. Earthworm casts are the excreta of earthworm. These casts contain five times the nitrogen of ordinary soil, seven times the phosphorus, eleven times the potash, two times the calcium and magnesium, and eight times the Actinomycetes (useful bacteria). During the passage through the earthworm's gut, organic materials are thoroughly shredded and mixed with mineral soil materials. Probably because of enhanced bacterial activity earthworm casts are usually high in polysaccharides, which are credited with stabilizing the granular structure. The earthworm castings are known to be a rich source of plant growth promoting substances such as Auxins and Cytokinins. The casting behavior of earthworms, therefore generally enhances the aggregate stability, exchangeable calcium and potassium of the soil.

3.9 VERMICOMPOST

Vermicompost is a result of a method of making compost with the use of earthworms, which generally live in soil. They eat biomass and excrete it in digested form. This compost is generally called Vermicompost . It is the dropping of earthworms after the intestinal digestion of organic matter. These dropping are high in nutritive value. Even of the Vermicompost dries there is no harm to the microorganisms. Research has revealed that Vermicompost contains many micronutrients like Manganese, Iron, Molybdnum, Boron, Copper and Zink as well as some of the growth regulators. Vermicompost is a stable fine granular organic matter, when added to clay soil loosens the soil and provides the passage for the entry of air. The mucus associated with the cast being hygroscopic absorbs water and prevents water logging and improves water-holding capacity. Thus in the sandy soil where there is problem of water retention the strong mucus coated aggregates of Vermicompost hold water for longer time.

In the Vermicompost some of the secretion of worms and the associated microbes act as growth promoters along with other nutrient. It improves physical chemical and biological properties of soil in the long run on repeated application. The organic carbon in Vermicompost releases the nutrient slowly and steadily into the system and enables the plant to absorb these nutrients. The nutrient level of the Vermicompost varies with the inputs. To get high nitrogen content residues of leguminous species should be added to the pit. Addition of blood meal will result in increased nitrogen and potassium content and bone meal will enhance the potash and phosphorus content of the Vermicompost. Vermicompost is richer than other types of compost. A comparative statement of Farmyard manure and Vermicompost is given in Table 1.

Table1: Comparative Statement of Farmyard Manure and Vermicompost

Sl.No	Particulars	FarmYard Manure	Vermicompost
1.	Nitrogen%	0.40–0.75	1.00–1.60
2.	Phosphorus%	0.17–0.30	0.50–5.04
3.	Potash%	0.20–0.55	0.80–1.50
4.	Calcium%	0.91	0.44
5.	Magnesium%	0.19	0.15
6.	Iron(ppm)	146.50	175.20
7.	Manganese (ppm)	69.00	96.51
8.	Zink (ppm)	14.50	24.43
9.	Copper (ppm)	2.80	4.89
10.	Carbon: Nitrogen Ratio	31.28	15.50
11.	Duration required for the preparation	One year	Less then ¼ th year
12.	Immunity against insect pest and diseases	Not developed	Developed

3.10 HOW TO COLLECT THE EARTHWORM?

About 500 gm jaggery and equal quantity of fresh cow dung should be mixed in 15to 20 liters of water, and this diluted slurry-should be sprinkled over the area. Wet pats of cow dung is scattered over the area and a layer of moistened rice straw should be laid over it. The whole area is then covered with a jute sack. Regular watering should continue for a period of 20 to 25 days and care should be taken to avoid waters stagnation. When the cover is removed a large worms can be seen. Farmers can collect these earthworms and utilize for Vermicomposting. A solution of formaline @ 0.55 % has been also found effective for the collection of Earthworms. Taking worms out of their natural environment and placing them in containers creates a human responsibility. They are living creatures with their own unique needs, so it is important to create and maintain a healthy habitat for them to do their work.

3.11 MATERIAS REQUIRED FOR VERMICOMPOSTING FOR A 10 SQ .M. PLOT

1	Stone chips of 1 cm size	For the filling of pit 3"
2	Sand or Morang	For the filling of pit 3"
3	Wet soil	For the filling of pit 6"
4	Dry organic matter	200-300 kg
5	Decomposed farmyard manure	300-400 kg
6	Organic waste including kitchen	700-800 kg waste
7	Earthworms	10,000
8	Water	Ready supply

3.12 METHOD OF PREPARATION OF VERMICOMPOST

Vermicomposting can be done indoors and outdoors, thus allowing year-round composting. Wooden plastic containers either build or buy, or something like an old dresser drawer, trunk or discarded barrel may be used for Vermicomposting. Wooden containers preferably should be used because it is more absorbent and a better insulator for the worms. In plastic containers compost tends to get quite wet. Containers should not be very large and heavy for easier lifting and moving. Depending on the size of the container drill 8 to 12 semicircular holes of ½ inches in the bottom for aeration and drainage. A plastic bin needs more drainage holes. Raise the container on bricks or wooden blocks and place a tray underneath to capture excess liquid, which can be used as liquid plant fertilizer.

The container needs a cover to conserve moisture and provide darkness for the worms. If the container is indoors, a sheet of dark plastic placed loosely on top of the bedding is sufficient as a cover. For outdoor containers, a solid lid should be preferred, to keep away-unwanted scavengers and rain. Worms need air to live, so be sure to have bin sufficiently ventilated.

It is necessary to provide damp bedding for the worms to live in, and to burry food waste in. Suitable bedding materials are cow dung slurry, shredded newspaper and cardboard, shredded fall leaves, chopped up straw and other dead plants, seaweed, sawdust, compost and aged manure. It is very important to moisten the dry bedding materials before putting them in the container. Do not use large size worms found in soil and compost, as they are not likely to survive. It is advisable not to compost meats, dairy products, oily foods and grains because of problems with smells, flies and rodents. No glass, plastic or tin foil should be present in the composting materials. Containers should be kept out of hot sun and heavy rain. If temperature drops below 40 degree F, containers should be replaced indoors or well insulated outdoors. It is estimated that 1000 tonnes of sludge organic waste could be converted in to 400 tonnes of organic fertilizer through vermicomposting. The production cost of vermicompost works out of Rs. 750 – 1000/- per tonne. A flow chart for the preparation of Vermicompost is given herewith.

FLOW CHART FOR THE PREPARATION OF VERMICOMPOST

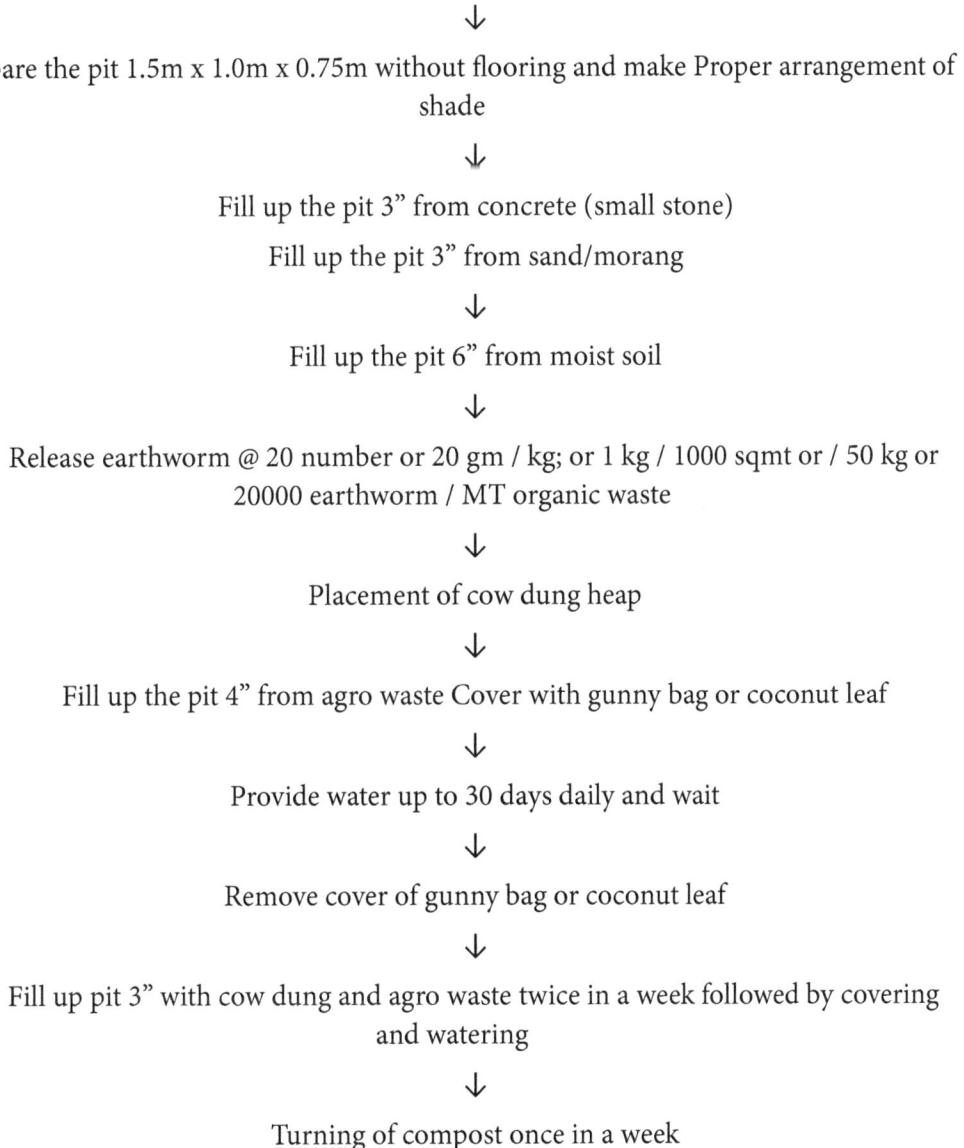

↓

Prepare the pit 1.5m x 1.0m x 0.75m without flooring and make Proper arrangement of shade

↓

Fill up the pit 3" from concrete (small stone)

Fill up the pit 3" from sand/morang

↓

Fill up the pit 6" from moist soil

↓

Release earthworm @ 20 number or 20 gm / kg; or 1 kg / 1000 sqmt or / 50 kg or 20000 earthworm / MT organic waste

↓

Placement of cow dung heap

↓

Fill up the pit 4" from agro waste Cover with gunny bag or coconut leaf

↓

Provide water up to 30 days daily and wait

↓

Remove cover of gunny bag or coconut leaf

↓

Fill up pit 3" with cow dung and agro waste twice in a week followed by covering and watering

↓

Turning of compost once in a week

↓

Continue the process of filling covering, watering and turning up to the completion of pit

↓

After complete filling of pit continue watering up to 45 days and turning once in a week

↓

Stopwateringfor2days

↓

Remove the compost from the pit

↓

Make a heap in shady place for 3days

↓

Collect the worm from the lower level of heap and again release them in the pit

↓

Sieving and packing of Vermicompost

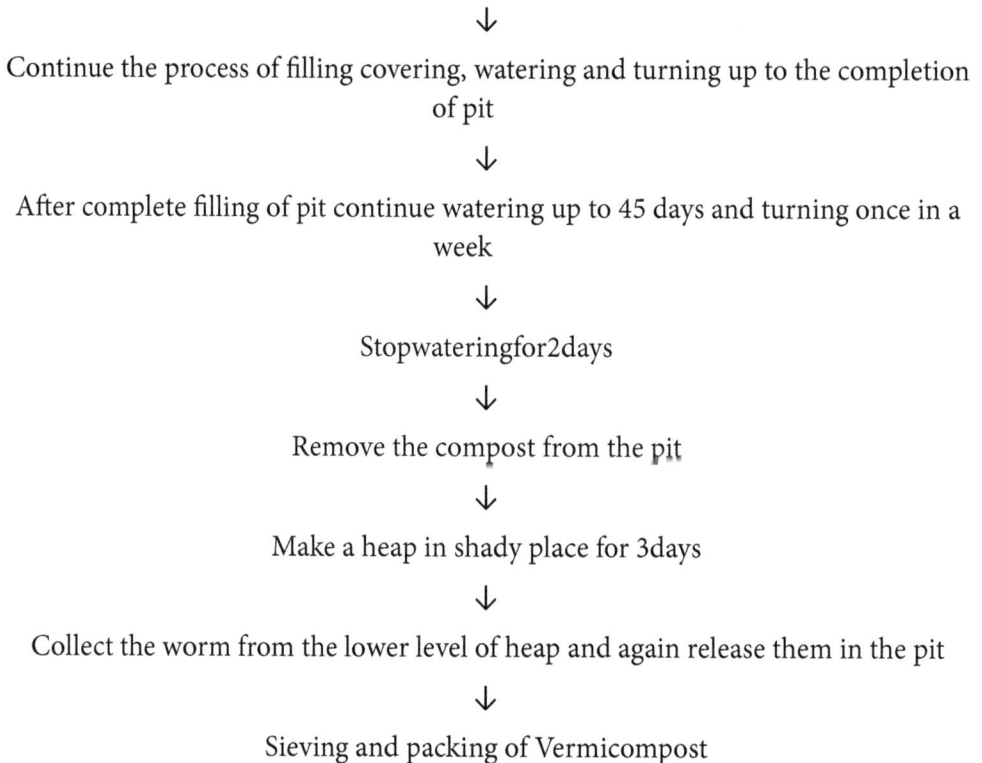

How to know that Vermicompost is ready?

⦿ The colour of compost turns black.

⦿ Compost will be light in weight.

⦿ Compost does not emit any foul smell.

⦿ pH value of the compost will be around seven.

⦿ The worms start crawling over the top cover and the bed boundaries.

3.13 MAINTENANCE OF VERMICOMPOSTING

Take care to maintain an optimum number of Earth worms in the pit/fields. Their population is adversely a ffected by-

1. Use of chemical fertilizers.
2. Use of certain pesticides against soil borne pests.
3. In appropriate cultivation techniques, like use of rotary cultivators.
4. Acidification of soil.
5. Insufficient organic matter in the soil.
6. Always maintain moisture 60 to 65%.
7. Make the heap in shady and comparatively higher site.
8. Worms have been known to crawl out of the bedding and on to the sides and

lid if conditions are wrong for them. If the moisture level seems all right, the bedding may be too acidic. This can happen if you add a lot of citrus peels and other acidic foods. Adjust by adding a little garden lime and cutting down on acidic wastes.

9. Worms require protection from excessive sunlight, heat, down pouring rain etc. They grow well under shades.

Protect the earthworms in the pit from their enemies like birds, rats, mice, toads,lizards,centipedes,antsand cockroaches etc.

3.14 QUANTITY OF VERMICOMPOST TO BE APPLIED

Sl. No	Type of crop	Quantity
1.	Rice, Wheat, Jowar, Bajra, Maize	2.50 tonnes/ha
2.	Cotton	3.75 tonnes/ha
3.	Groundnut, Mustard and Pulses	2.50 tonnes/ha
4.	Sugarcane	5.00 tonnes/ha
5.	Potato, Tomato, Brinjal, Carrot, Cauliflower, CabbageandGarlic	1.00–2.00 tonnes/ha.
6.	Coconut and Mango	4-5kg/plant (below 5 years)
		8-10 kg/plant (above 5 years)
7.	Lime and Pomegranate	3-4 kg/plant (below 5 years)
		6-8 kg/plant (above 5 years)
8.	Pumpkin, Papaya, Orange, Pear and Peach etc.	6-8 kg /plant
9.	Rose, Jasmine, Marigold etc.	3.75 tonnes /ha or 2-3 kg/plant
10.	Chilly, Turmeric, Ginger	3.75 tonnes /ha
11.	Grapes, Pineapple, Banana	3.75-5.00 tonnes/ha
12.	Plants in pots	250 gm/pot

3.15 PROCESSING: TIME AND ACCELERATION

1. It is possible to get vermicompost in 5 to 6 weeks with high worm populations and frequent management; 2 to 3 months (60 to 90 days) under favourable conditions; but 4 to 6 months is a better estimate with minimal management of the worm beds.

2. With high volume flow through systems, it has been reported that a marker such as a coin placed on the surface of the bed will typically drop out the bottom of the bed in about 60 days.

3. With a high worm population, kitchen wastes or animal manure will be decomposed in 4 to 6 weeks. If the material is to be used in certified organic production systems, the required worm composting time for a batch system is four months (16 weeks).

4. Worm populations will clearly decline with no added feed for four months. Worms are extracted over a one-month period. A low population of worms is

left for up to one month prior to sieving the finished compost. This also allows time for additional worms to emerge from cocoons.

3.16 MATURITY AND STABILITY

Compost/vermicompost quality is assessed on the basis of its stability and maturity. Good compost would have the texture of moist loose soil homogeneous and aesthetic. The abundance of physical, chemical and biological changes occurred during aerobic or worm composting. Different parameters proposed to assess the maturity of the compost include the C/N ratios, water soluble carbon, cation ion exchange capacity, CO_2 evaluation, NH4-N/NO3- N ratio, organic carbon content, and humus content. However, germination index (GI) measuring phototoxicity has been considered as a reliable parameter to quantify compost maturity. A coliform test gives indication of pathogen reduction.

3.17 COMPOSTING AND VERMICOMPOSTING A COMPARISON

In contrast to composting, vermiculture has several distinct applications, with the potential to produce different grades of end product, depending on volume or time constraints:

1. The complete processing of organic wastes by this method produces the highestgrade end product, in the form of worm casts. These typically contain much higher concentrations of vital nutrients than standard composted material. Worm casts tend to be used as a high quality (and high value) soil conditioner within the horticultural sector, rather than as bulk compost or plant bedding material.
2. The partial processing of organic material, in order to accelerate the composting process or to provide a product of higher quality than standard compost.
3. Elimination of nuisance odours associated with the decay of organic matter, such as in forms of open air composting, which do not employ sealed 'in-vessel' equipment.
4. The energy requirements of vermicomposting are very small compared to the existing waste disposal systems and processing costs are negligible.
5. The breeding of worms. Although this is not of primary concern for a municipal composting installation, such a facility would require very large numbers of worms in order to operate satisfactorily. The maintenance and increase of worm numbers is therefore necessary, in order to increase initial worm numbers as the facility expands.

3.18 PROCESS ACCELERATION

3.18.1 . Using Organic nutrients and other additives.

Literature survey on this area of research showed that only very few reports were available on the successful use of organic nutrients or other additives for enhancing the vermicomposting process. A recent article by Parray, et al, 2014 reported the use of Spirulina and Trichoderma as probiotic and microbial inoculants during pre-decomposition period in order to get qualitative and quantitative improvement vermicomposting. Studies by Vasanthi,et al, 2011 recommended the use of an organic nutrient, Jeevamirtham (a preparation using cow dung, urine, jaggery and black gram flour) for vermicomposting to enhance the functioning of the earthworm and to increase fertilizer value of vermicompost.

3.18.2 . Using Effective Microorganisms(EM)

EM is a multi-culture of co-existing anaerobic and aerobic beneficial microorganisms. (Higa, 1991;Fig.10).The major groups of the microbes present in EM are:

- ⊙ Lactic acid bacteria - Lactobacillus plantarum,L.casei , Streptococcus lactis.
- ⊙ Photosynthetic bacteria - Rhodopseudomonaspalustrus, Rhodobacterspaeroides.
- ⊙ Yeasts - Saccharomyces cerevisiae, Candida utilis.
- ⊙ Acitinomycetes - Streptomyces albur, S.griseus.
- ⊙ Fermenting fungi - Aspergillus oryzae, Mucor hiemalis.

They are used directly in waste management programmes as they can grow and multiply in solid wastes and other residues under proper conditions and are capable of converting wastes into high-quality compost. Studies by the author in 2013 showed that EM can be used to convert different types of lignocellulosic residues from a large wood industrial complex into a reusable form (E.Sreenivasan, 2013b). Research works by this author, with the objective of enhancing the efficiency of the earthworm that is involved in the process of vermicomposting of wood waste, by fortifying the vermibed using an effective microbial suspension were successful during the initial trial. This work requires further experimentations to establish and recommend the utilization of Effective Microorganisms for enhanced functioning of the earthworms.

3.18.3 Advantage of Vermicompost

1.	Vermicompost boosts grow thof plants making them strong and healthy and free from pest attack. Ithelps micro organisms produce polysaccharides, improving soil health.
2.	Vermicompost absorbs 10 times more water than the soil, soit increases water retention capacity, thus avoiding erosion. Earthworm plough through 7-8 times a day, making the soil loamy, thereby enhancing drainage. They alsotakein10 times their body weight in water and release it to the soil when needed.
3.	Vermicompostorganicelementsdecomposesofinelyinto0.2-micronsizeintheearthworm'sstom achenablingrootsofplantstoeasilyabsorbthesefood elements.
4.	Vermicompost contains abundant food elements; micro bacteria and humus,making it complete manure. Using Vermicompost thus not only saves money, but also increases yield by 40% to 80% per hectare.
5.	Vermicompost increases the shape, colour, taste and lusterof crops. It increases shelf life andnutritions of fruits,vegetables, cereals and flowers.
6.	Vermicompost is about 50% cheap erth an chemical fertilizers, saving

	Expensive imports. This will give support to export of chemical and pollution free produce to foreign countries there by earning valuable foreign exchange to the country.
7.	Vermicompost increase immunity to crops. Therefore, no money is spent on unaffordable chemical fertilizers, pesticides and insecticides. Consuming fruits, vegetables, and grains, grown thus is safe and everyone is free from health hazards.
8.	Vermicompost is suitable for all type of soil, crops and can be used in any season.
9.	Vermicompost is rich in several micro floras like Azospirillium, Actinomycetes, and Phosphobacillus, which multiply faster through digestive system of earthworms.
10.	Several enzymes, auxins and complex growth regulators like Gibberellines, which are not formed in different soils and environmental conditions, are present in the earthworm castings.
11.	Buffering action neutralize soil pH.
12.	Vermicompost helps multiplication of earthworms, which reduce the incidence of nematodes
13.	Due to buffering action, minerals and trace elements become available more easily to crops.
14.	Leaching of nutrients from chemical fertilizers in the soil is reduced considerably specially of Nitrogenous fertilizers.
15.	Reduce soil toxicity by buffering action.
16.	Vermicompost influences the physiochemical and biological properties of soil where in turn improves soil fertility.
17.	Low cost to produce in comparison of fertilizers.
18.	Easy to use in comparison fertilizers.
19.	Harmless to useful soil organisms
20.	Converts organic matter to useful plant food.
21.	The fresh Vermicompost will have maximum microbial load beneficial to increase soil microflora.

3.19 ADVANTAGE OF VERMICOMPOSTING

- Productive utlization of waste materials available in or around the house on farm.
- Assured supply of Vermicompost which is excellent in manurialqualitiesandas soil conditioners.
- The waste biomassremaining after harvesting the crop can be recycledintothe same field thus assuring maintenance of soilfertility.
- Transportation cost of manure can be avoided.
- Vermicompost preparedat site will avoid loss of neonates (newly bornearthworms) andcocoons.
- Viability of earthworm cocoons and their number is guaranteed when madeatsite.
- This operation will stimulate population growth of local communities of earthworms, imparting a multiplier effect, which in turn, will help in maintaining so ilfe rtilityand ecology.
- By manipulating composition of wastes, desired quality of Vermicompostcan be obtained.

⊙ Vermicomposting process generates an internal heat. The heat kills the pestsand pathogens. It also destroys the seeds of weeds that may be found in the organic wastes.

Nitrogen oxides from chemical fertilizers might deplete the Ozone layer which mayconsequently cause-more skincancer, more eye cataracts, loss of immunity, harm of the phytoplankton and crop damage affecting entirefood chain but Vermicomposting will help for the minimization of thesecosequences.Around the world, there is growing interest in finding alternatives to the industrial farming methods that have emerged during the 20th century. Almost five decades of pesticides and fertilizers use have left us a tragic legacy; severe contamination of our soils and water system, increased cancers, birth defects and other ailments in humans and the emergence of powerful pests which are resistant to chemical pesticides. According to the World Health Organization (WHO) approximately one to two million persons are affected every year because of pesticides. The present level of use of chemical fertilizers and pesticides is harmful to ecological balance and sustainable use of natural resources. Vermiculture biotechnology can provide effective alternative to these costlier agricultural inputs while preserving the environment. During the formulation of development programme all developmental agencies are required to pay attention that there should be no any adverse effect on the environment by the implementation of development programme. Promotion of Eco-friendly agriculture is certainly saving the earth itself as well as saving of own self. It is important to conserve environment for regular development and welfare of humanity on our unique planet earth.

3.20 VERMICOMPOSTING SYSTEMS

In general the following vermicomposting systems are used the world over, for volume reduction, extraction of organic load, cost and energy reduction and rapid processing. Any of these systems may be adopted for vermicomposting depending on the availability of space, nature of waste or bedding material, quantity of waste to be processed etc.

3.20.1 Windrow System

This system deals with construction of windrows under shade to avoid direct sunlight. This method involves spreading out a layer of worms and bedding on the floor or ground to start(Fig.6).The first layer of a new windrow should be 10 to 15cm high. Earthworms can be reared at a production nursery or rectangular boxes prior to their inoculation in the windrows. The worms feed from the bottom till the top of the bed. The windrow has to be monitored daily and when signs of surface feeding are noticed, another 7 to 10 cm layer of feedstock can be added. Thick layers of feedstock are avoided because they impede oxygen penetration into the windrow. This can cause the worms to migrate to the upper surface before lower layers are thoroughly digested, creating anaerobic fermentation. The windrows are irrigated with center post sprinkler up to twice daily to maintain optimum

Excessive Heating

Another of the challenges to any vermicomposting system, irrespective of size, is the potential for heating in the feedstock. Bacteria are the primary decomposers of raw organic matter and in oxygen rich system, water, carbon dioxide and heat are produced as a result of microbial activity. When raw material is added to the system, particularly in large volumes, the mass can support the activity of billions of bacteria. Bacterial activity can produce significant amounts of heat, which may be trapped within the system. Even a small volume of raw material can result in heating if it contains sufficient energy to support high levels of bacterial activity. This potential for heating complicates the assessment system loading rates. It should be recognized that a worm bed may contain thousands of different species of invertebrates and microorganisms, all of which play a vital role within the vermiculture ecosystem. The loading rate cannot therefore be based solely on the needs or capacity of a single organism in that system. Bacterial activity may have as much impact as the worm activity, as bacteria will have access to the feedstock first.

Overfeeding (in relation to the design capacity, the type of feedstock and/or the level of system activity) may generate sufficient heat to deter worm activity. Unless design modifications can be made, such as installing fans to remove excess heat, the loading rate will need to be decreased to a point where heating is not a problem, even if that means feeding less material than the worms are capable of processing.

moisture content of 80% throughout the windrow. Now the material on top is removed to start a new row, and the material on the bottom is vermicompost. The new row can be started in a new location. Or it can be moved longitudinally by about 20' by dumping the worm inhabited material past the end of the row, digging out some vermicompost, and then shuffling some more worms over. One of the problems with this method is that it requires some digging by hand. If the windrow is wide enough, it might be possible to drive into the side of it with a loader to remove some material off the top. But the rest of it will have to be forked off by hand because the loader will merely push it over and off the opposite edge of the row. No matter how careful you are, material will roll down the sides when digging off the top, and it takes some care to make sure that material does not end up getting mixed in with the finished product. Windrows work great in places where temperatures are just right most of the time. If the windrows are outside, they should ideally be under a shade structure on a slightly elevated concrete slab. If there is no shade structure, or if the shade structure does not shed rain, then compost covers need to be ready for use during heavy or frequent rain. Compost covers are heavy and awkward, usually requiring two or more people to handle. If there is no concrete, there still needs to be some type of impervious surface such as asphalt or certain types of clay.

Fig.6: Vermicomposting windrows

3.20.2 Wedge System

This is a modified windrow system and it maximizes space and simplifies harvesting because there is no need to separate worms from vermicompost. Organic wastes are applied in layers against a finished windrow at a 45 o angle. The piles can be constructed inside a structure or outdoors if they are covered with a tarpaulin or compost cover to prevent leaching of nutrients. A front-end loader is used to establish a windrow 1.2 to 3 m wide by whatever length is appropriate. Spreading a 30 to 45 cm layer of organic materials the length of one end of available space starts the windrow. Up to 0.45 kg of worms is added per square meter of windrow surface area. Subsequent layers of 5 to 7.5 cm of organics are added weekly and preferably more addition in colder seasons. After the windrow reaches two to three feet deep, worms in the first windrow will eventually migrate toward the fresh feed. Worms will continue to move laterally through the windrows. After two to six months, the first windrow and each subsequent pile can be harvested. A variation of this method uses a migrating windrow. Here a row can be started as a windrow or as a layer and then grown into a windrow. A loader is used to feed one side of the row, keeping it the same height and length, but making it wider. After a while, a loader is used to remove vermicompost from the side opposite the one being fed. Later, the finished compost is removed from that same side again. The row migrates laterally as it is being fed along one side and harvested along the other. At any time, the feeding and harvesting sides can be switched to change the migration direction.

3.20.3 Container System

a. Pits, Tanks and Cement rings

Pits made for vermicomposting are 1 m deep and 1.5 m wide. The length varies as required. Tanks made up of different materials such as normal bricks, hollow bricks, asbestos sheets and locally available rocks were evaluated for vermicompost preparation. Tanks can be constructed with the dimensions suitable for operations. Scientists at ICRISAT have evaluated tanks with dimensions of 1.5 m (5 feet) width, 4.5 m (15 feet) length and 0.9 m (3 feet) height. The commercial biodigester contains a partition wall with small holes to facilitate easy movement of earthworms from one tank to the other (Nagavallemma, et al, 2004)(Fig.7)

Fig.7: Vermicomposting tanks made of ordinary bricks in a semi-commercial unit

Vermicompost can also be prepared above the ground by using cement rings. The size of the cement ring should be 90 cm in diameter and 30 cm in height.

b. Commercial model

The commercial model for vermicomposting consists of four chambers enclosed by a wall (1.5 m width, 4.5 m length and 0.9 m height). The walls are made up of different materials such as normal bricks, hollow bricks, asbestos sheets and locally available rocks. This model contains partition walls with small holes to facilitate easy movement of earthworms from one chamber to another. Providing an outlet at one corner of each chamber with as light slope facilitates collection of excess water, which is reused later or used as earthworm leachate on crop. The four components of a tank are filled with plant residues one after another. The first chamber is filled layer by layer along with cow dung and then earthworms are released. Then the second chamber is filled layer by layer. Once the contents in the first chamber are processed the earthworms move to chamber 2, which is already filled and ready for earthworms. This facilitates harvesting of decomposed material from the first chamber and also saves labor for harvesting and introducing earthworms. This technology reduces labour cost and saves water as well as time.

Fig.9: A view of a semi-commercial vermicomposting unit

c. Beds or Bins

1 Top-Fed Type

A top-fed bed works within four walls and (usually) a floor, often within a building. If the bins are fairly large, they are sheltered from the wind, and the feedstock is reasonably high in nitrogen, the only insulation required may be an insulating "pillow" or layer on top. These can be as simple as bags or bales of straw. The reader should note that these beds were designed for vermiculture, rather than vermicomposting. Harvesting vermicompost can be most easily accomplished by taking advantage of horizontal migration. To harvest, the operator simply stops feeding one of the beds for several weeks, allowing the worm's time to finish that material and then migrate to the other beds in search of fresh feed. The "cured" bed is then emptied and refilled with bedding, after which feeding is resumed. This is repeated on a regular rotating basis. If the beds are large enough, they can be emptied with a tractor instead of by hand.

2. Stacked Type

One of the major disadvantages of the bed or bin system is the amount of surface area required. While this is also true of the windrow and wedge systems, they are outdoors, where space is not as expensive as it is under cover. Growing worms indoors or even within an unheated shelter is an expensive proposition if nothing is done to address this issue. Stacked bins address the issue of space by adding the vertical dimension to vermi- composting. The bins must be small enough to be lifted, either by hand or with a forklift, when they are full of wet material. They can be fed continuously, but this involves handling them on a regular basis. The more economical route to take is to use a batch process, where the material is premixed and placed in the bin, worms are added, and the bin is stacked for a pre-determined length of time and then emptied. This method is used by a number of professional vermicompost producers.

The initial cost of setting-up a stacked-bin system is high. It requires a shelter, bins, a way to mix the bedding and feed, and equipment to stack the bins, such as a forklift. On a smaller scale, of course, this could all be done by hand. Another disadvantage arises when it comes time to harvest. As with the batch windrow systems, the worms are mixed in with the product and need to be separated. That requires either a harvester or another step in the process, where the material is piled so that the worms can migrate into new material.

d. Continuous Flow Systems

This system originally developed by Dr.Clive Edwards of the Rothamstead Agricultural Research Station is gaining popularity and has been adopted by many mid-scale operations. The efficiency savings offered by their continuous flow design increases with the amount of material processed. This system design is now almost ubiquitous in commercial mid to large-scale vermicomposting systems. Each of these systems uses a relatively deep top-fed container, in which the composting mass sits upon a raised floor made from a widely spaced wire mesh. Worms are added to the system and food waste is added gradually, layered with bedding material. The system is continually fed until the bin is nearly full. The worms generally move upward through the feedstock/bedding layers and vermicompost is harvested from below by scraping or cutting a thin layer of finished material from just above the grill using a rake or a manually or hydraulically-operated blade. Continuous flow systems offer several advantages to medium to large scale composting operations. They are relatively straightforward to construct and operate. They are labourefficient in terms of operation and harvesting finished material. They avoid the need for expensive equipment associated with technical 'in-vessel' systems and the turning and screening of windrowed material. It should be noted that, despite the recent and increasing interest in this design, windrows are still the most common large-scale vermicomposting system in use. Continuous flow vermicomposting designs are arguably the most efficient systems available, in terms of time and labour savings. However, regardless of efficiency or ease of operation, there is no design that eliminates the need for careful monitoring and good system management, which may require considerable initial experimentation and familiarization.

Maintaining Continuous Flow

Continuous flow vermicomposting systems provide an ongoing flow of vermicompost that is easily removed from the system without disrupting the worm activity or requiring complex or time-consuming harvesting methods. Because of their operating efficiency, these system designs are becoming almost as popular as windrows for large-scale applications. However, like all vermicomposting systems, the continuous flow model poses several challenges. In order to simplify some of the technical terminology, the worms most often used in vermicomposting are usually referred to as "surface feeders." They are generally presumed to only be active at or just below the surface. However, this is not always the case. Earthworms are oxygen breathing, moisture-loving animals that require organic material to be bacterially active before they eat it. In their natural environment, this is usually top few inches of soil or surface organic litter, such as leaves. In any system with a free flow of oxygen, monitored moisture level and abundant supply of decomposing organic material, earthworms may spread throughout the material unless the system is carefully managed. Earthworms may therefore be found anywhere within the continuous flow systems which meets their requirements.

One of the advantages to the continuous flow design is in the ease with which a continuous supply of vermicompost can be removed from the system. However, harvesting of the finished material should not begin until the system is nearly full of material. Many operators have found that, along with appropriate loading rates, a minimum depth of material in the system of between 12"-18" will help to ensure that few, if any, worms will be low in the bed and drop through, or fall out with the harvested vermicompost. Once fully charged, vermicompost then needs to be removed at a rate that maintains a relatively constant level of material in the system.

Feeding Rates

The precise loading rate (at which raw feedstock can be added to a worm bed to encourage the worms to concentrate at or near the surface) will vary depending on the feedstock being used, temperature, moisture levels and the density of the worm population. Proper loading rates require that new feedstock is not added until the majority of the previously added feedstock has been decomposed. Adding new feedstock too early means there can lead to a build-up of unprocessed material within lower layers. There will therefore be sufficient available food deeper within the container, instead of being concentrated immediately below the surface. The worms will then spread into all the available food areas. Worm movement in the lower levels of a flow-through system often causes vermicompost to drop through the mesh floor before it has been sufficiently decomposed. Also, when the system is harvested, worms remaining low in the material will fall through with the vermicompost and will either need to be separated using labour-intensive screening methods, or will be lost to the system. Most operators of continuous flow systems find that frequent additions of thin layers of feedstock (1"-2" deep spread across the surface) produce the best results. Feedstock is sometimes mixed with bulking agents like compost, shredded leaves, cardboard, paper or straw, or covered with an equally thin layer of these materials. Paper products are a preferred feedstock for earthworms, as they provide an easily accessible and digestible form of carbon.

4

FACTORS AFFECTING THE COMPOSTING PROCESS

4.1 GENERAL

The management of the composting process depends on the periodical control of temperature, moisture content and oxygen concentration. The determination of the quantities of ammonium (NH_4-N), nitrite (NO_2-N) and nitrate (NO_3-N) in the compost allows the evaluation of its maturity, the quality of the composting process and of the storage of the product, and the risk of N immobilization or release in the soil. Plant phytotoxicity tests can assess the maturity and the quality of the produced compost.

The choice of the appropriate compost, application strategy and use, should be guided by analyses of pH value, salinity, intensity of colour extract and nutrient content. Certain environmental factors affect the composting process, so they require food (carbon and nitrogen), air, and water to make compost. Along with this, they also require a favourable temperature and pH for rapid composting. Other physical factors affecting the pace of composting include surface area, particle size and volume. These factors affecting the process are detailed as below.

4.1.1 Organism

For every natural occurring substance, there is a corresponding microbial enzyme complex with the ability to convert it to carbon dioxide, humic material and waste. Organic waste usually is heavily contaminated with its population of micro-organisms, i.e. bacteria, fungi and actinomycetes which are derived from the atmosphere, water or soil. Micro-organisms also need the supply of nutrients, air, water and favourable conditions to work effectively. These micro-organisms can work effectively if the get large surface area to act and some macro-organisms, *i.e.* mites, centipedes, snails, millipedes, springtails, spiders, slugs, beetles, ants, flies, nematodes, flatworms, rotifers, and earthworms etc. play an important role in cutting, tearing and chopping the organic substrate so the micro-organisms may find suitable to act upon. Presence of a sufficient number of these assures quality compost production.

The culture of cellulolytic organisms did not compete well with the native microflora during the composting of the waste. The inoculation with the one-litre slurry containing 5% each of dung, soil and well-decomposed compost is enough to introduce the microflora which can increase the decomposition rate and may bring maturation of the compost quickly.

4.1.2 Organic Waste

Organic wastes are the primary substrate subjected for decomposition. It provides food for organisms in the form of carbon and nitrogen. As stated earlier, bacteria use carbon for energy and nitrogen to build protein to grow and reproduce. The level of carbon and nitrogen differs with kind of substrate. Organic materials rich in carbon tend to be dry and brown such as plant leaves, straw and wood chips. Nitrogenous materials tend to be wet and green such as fresh grass clippings, products of animal origin and food waste. There is a simple thumb rule to estimate carbon/ nitrogen content of organic matter. Any fresh and juicy materials will be usually higher in nitrogen, and older, drier, and woodier plant material will be higher in carbon.

The organic material with higher nitrogen will decompose faster than those, which are high in carbon content.

A C:N ratio between 25:1 to 30:1 is the optimum combination for rapid decomposition of organic matter. If the ratio is more than 30:1 that indicates high carbon and this limits the growth and reproduction of micro-organisms thus number becomes insufficient to produce heat and decomposition slows. It happens when the pile has more of dry leaves or wood chips, and decomposition will take a year or so. When C:N ratio is less than 25:1 this indicates too much of nitrogen and pile is likely to utilise excess nitrogen in the degenerative process and causes release malodorous ammonia gas. Too much nitrogen (N) may also cause a rise in the pH level leading to putrefaction. Any of the organic material may not have the desired C:N ratio but the blending of different materials will help in achieving a satisfactory C:N ratio. A volume-based simple thumb rule of using one-fourth to half of the organic matter as nitrogen-rich and remaining high carbon-containing materials like dry leaves and straws may be adopted for the purpose. It will provide adequate sources of carbon and nitrogen to micro-organisms for growth and cell division. C:N ratio of some of the common organic wastes used for composting is given in Table 1.

Table 1: Carbon: Nitrogen (C:N) Ratios of some common organic wastes

MATERIAL	C:NRATIO
Maizes talks	50-100:1
Straws of paddy, wheat, millets, etc.	40-100:1
Sugarcaneleavestrashand baggage	50-100:1
Pigeonpeastalks	50-100:1
Oldgreen leaves andtwigs of plants	50-100:1
Mature grasses	40-80:1
Coconut fibrewaste	50-100:1
Peanut hulls	40-80:1
Mustard plants (afterharvest)	40-80:1
Paper wastes	170-200:1
Sawdust	200-500:1
Silk mill wastes	30-50:1
Potato waste	40-80:1
Fruit waste	30-35:1
Grass clippings	12-25:1
Hay, green	25-30:1
Fresh leaves from elderandelm	21-28:1
Fresh Leaves (Pine)	60-100:1
Fresh Leaves, other	30-80:1
Manureofcattle,buffalo,pigandpoultry	20-25:1
Vegetable waste	12-25:1
Weeds	25-30:1
Woodchips	500-700:1

The nitrogen (N) in composts is mainly present as organic nitrogen, which is less available to plants. However, the greatest proportion of N taken up by plants is in the form of mineral-N. Three forms of mineral nitrogen are relevant in compost: ammonia (NH_4-N), nitrite (NO_2-N) and nitrate (NO_3-N). The concentration of these three forms is evaluated during the composting process. NH_4-N is the first form of mineralized nitrogen found in compost when the organic material is decomposed. NH_4-N is soluble in water and when the moisture content becomes too low, the NH_4-N is lost as it will be formed into gaseous NH_3 (ammonia). NO_3-N. During the curing process, nitrification is ongoing and the NH_4-N is transformed into NO_3-N. If oxygen starvation happens during the curing phase or the storage, bacteria can use the oxygen of NO_3 and transform it back to nitrite (NO_2; toxic for the plants) or to nitrous oxide (N_2O; strong greenhouse gas). NO_2-N is an intermediate, phytotoxic product arising during the nitrification. It can also be a result of the denitrification process by oxygen starvation at the end of the curing process or from compost storage.

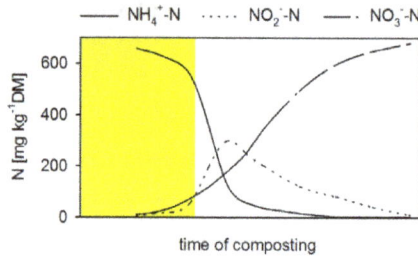

Fig. 1: Evolution of the mineralized forms of nitrogen in the compost piles during the composting process

If the composting mixture is carbon-rich or when the composting process is inappropriately managed (*e.g.*, low moisture content), it is possible to find almost no mineral N in the compost. The N is immobilized in the microbial community or is lost as NH_3. In this case, the composting process can become blocked because of lack of available N. When such compost is applied, it is also possible that N immobilization happens in field, and the plant growth can then be inhibited if no other sources of nitrogen are added. So, the values of mineralized forms of N are essential parameters to management of the composting process on one hand and the identification of appropriate uses for compost products on the other. NH_4-N, NO_2-N and NO_3-N can be easily analysed with different quick tests in 0.01 M $CaCl_2$ or KCl compost extracts. The interpretation of the obtained measurements is described in Table2. It is not the absolute individual values that are the most important, but the relationships between the different forms of mineralized nitrogen.

Table 2 Interpretation of the signification of the quantiy of the different forms of mineralized nitrogen forms in compost.

Presence of the N_{min} form[1]			Interpretation
NH_4-N	NO_2-N	NO_3-N	
-	-	-	No available N. Mixture too rich in carbon, or all NH_4-N was lost because of lack of moisture. If the compost is carbon rich: risk of nitrogen immobilization in the field. Recommendation: mix some N-rich material to the mixture (digestate, lawn, chicken litter, etc.).
++ / +++	-	-	Young compost (or digestate). Nitrification has still not started. Recommendation: keep the mixture moist enough to avoid NH_4-N losses and allow nitrification.
++/+++	++	+ / ++	Nitrification process starting. Recommendations: keep the mixture sufficiently moist to avoid NH_4-N losses; make sure that the oxygen supply to the mixture is constantly sufficient
+	+/++	++/+++	Nitrification process is progressing. Recommendation: make sure that the oxygen supply to the mixture is constantly sufficient
-	-	++/+++	Nitrification process achieved. Recommendation: make sure that the oxygen supply in the mixture is constantly sufficient Compost is mature and ready to be used.
-	++/+++-	++	Oxygen starvation problem. Recommendation: improved aeration of the compost.

1 -: none (< 10 mg N / kg DM); +: low quantity (10-50 mg N / kg DM); ++: medium quantity (50-200 mg N / kg DM);
+++: high quantity (> 200 mg N / kg DM)

4.1.3 Aeration

Aeration refers to the amount of oxygen in the system, and it is the key environmental factor. Organisms present in the compost pile can degrade organic materials either aerobically or anaerobically. Many organisms including aerobic bacteria need oxygen to produce energy, grow and reproduce. The types of organisms active in the pile and the metabolic process used to degrade organic compounds are related to the oxygen content of the system. Aerobic degradation is preferred for rapid composting.

Aeration in a compost pile occurs naturally as the oxygen-deficient air present in a pile warmed up due to composting, it leaves the pile and replaced by the fresh air from the surrounding. The process of aeration can also be affected by wind, moisture content, and porosity (spaces between particles in the compost pile). Compaction occurs with the progression of decomposition, which reduces the porosity of the piled mass. Further, the higher ratio of fine organic material like pine needles, grass clippings, or sawdust in the substrate reduces the porosity. Air circulation can also be impeded if materials become water saturated. If poor aeration is observed then turning of the material with shovel helps in better air movement, and it also increases porosity.

4.1.4 Moisture Content

Moisture influences the rate of the microbial population which helps in faster and proper degradation of composted matter. Maintaining adequate moisture content is essential since it provides the humidity required by micro-organisms for optimal degradation. The moisture content of 50-60 per cent is generally considered optimum for composting. Microbial induced decomposition occurs most rapidly in the thin liquid films present on the surfaces of the organic particles because water dissolves the organic and inorganic nutrients present in a pile and make them available for utilisation by micro-organisms. Too little moisture (<30%) inhibits bacterial activity, whereas too much moisture (>65%) results in slow and anaerobic decomposition causing odour production and leaching of nutrient. Composting should be carried out underneath some cover to control moisture level.

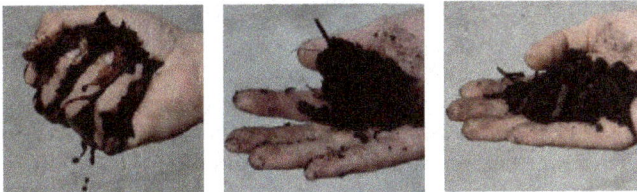

Fig. 2: Fist test to control the humidity of compost during the composting process. From left to right: too wet, optimal, and too dry.

4.1.5 pH

The pH of most of the composting substrate is slightly acidic, i.e. 6.0. At the early stage there is a production of organic acid hence pH goes again acidic 4.5-5.0. As the decomposition processget over and temperaturereduces and pH of the composted mass start increasing. It gets converted into alkaline pH 7.5-8.5 from acidic pH.The pH of the mature compost is 7.5-8.5.

4.1.6 Temperature

Temperature is another important factor in the composting process and is related to proper air and moisture levels. As aerobic decomposition is an oxidative process, so considerable heat is generateddue to microbial activity, which in turn increases pile temperature. Soil micro-organisms are metabolically active over defined temperature ranges. With the increase in the temperature of the pile, different groups of organisms become active. The temperature of the compost pile with substrates of appropriate particle size may rise to 65-75^0 C if oxygen, moisture, carbon and nitrogen is present in ample amount.Temperatures in the range of 32-60^0 C are typical of a well-operated system and are indicative of rapid composting. Higher temperatures begin to limit the microbial activity and temperature beyond 70^0 C becomes lethal to most micro-organisms. At this temperature, most of the weed seeds, insect larvae and potential plant or human pathogens that may be existing in the composting materials get destroyed. Although, composting will occur without careful temperature control, but maintaining thetemperature around 32-60^0 C is necessary for rapid composting.

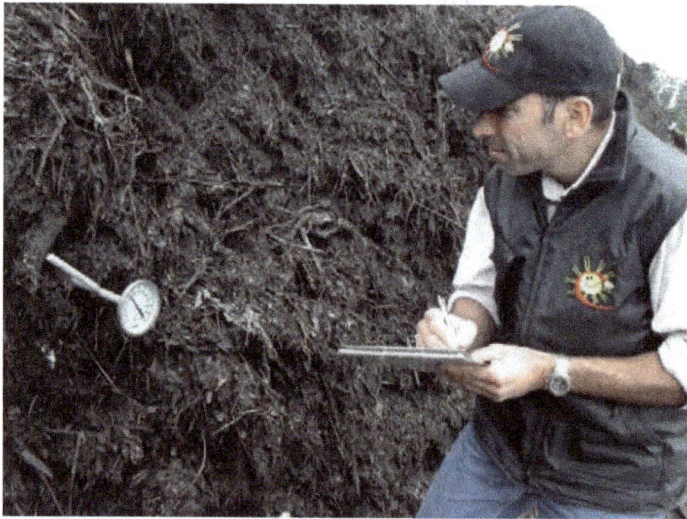

Fig. 3: Temperature measurement is important to control the composting process.

4.1.7 Surface Area

All the microbial activity is performedon the surface of the particle, so thesurface area of the organic material exposed to soil organisms influences the rate of decomposition greatly. Composting materials should be shredded, chopped or otherwise reduced in size to increase the surface area to increase the rate of decomposition. However, on the other hand, when particles are reduced to too small and compact, air circulation through the pile is inhibited. It decreases available O2 to micro-organisms within the pile and ultimately decreases the rate of microbial activity.

4.1.8. Size and Shape of Compost System

Size is a factor in retaining compost pile heat. A compost pile must be of adequatesize to prevent faster dissipation of heat and moisture, but it may be small enough to allow good air circulation. A pile of one cubic meter is an ideal size however size largely depends upon the method of composting. Smaller composting piles will decompose the material, but there may not be sufficient heat to destroy weed seeds and kill the germs as well, and decomposition is likely to take longer time.

The shape of the pile helps in controlling its moisture content. Outdoor compost systems may be sheltered from precipitation in humid regions; whereas, in arid regions, piles with a concave top is preferred to catch precipitation and any other added water.

The shape of the compost pile helps in regulating moisture content. In most humid and temperate climates, an egg-shaped or pyramidal (triangular) shape pile will work better. Even in higher rainfall zone sheltering from precipitation is desired. If it is a dry climate, then cutting the tip-off of the pyramid and making an indentation to catch precipitation may be desired. If it is too dry, then using pit will be better option to maintain proper moisture. Decomposition process halts as the pile dries and this kills all the organisms.

4.1.9 Time of Composting

The best times to build the compost pile is autumn and spring. In the autumn, many of the weeds and grasses will be flowering or started to go to seed so substrate of much higher C/N ratio will be available. In the spring, there will be again fresh green growth, which will have a lot more nitrogen (N). If the pile is built in mid-Summer, then it is likely to get too hot, and enough amount of organic matter may be lost, or if built in mid-Winter, it is more likely to stay too cold/wet and may result in anaerobic that may result in leaching of valuable nitrogen. In the summer, keeping the pile in the shade and adding more carbonaceous materials helps in keeping it cool, while in the winter making the pile on the south-facing side of a building and keeping the pilemay keep it warm.

4.1.10 Maturity Level

Several parameters give information on the maturity level of the compost. Principally, mature compost is stable and biologically less active compared to a young one. This means that the temperature of the compost does not rise anymore after compost turning, and also that the respiratory activity is low. The C/N ratio of mature compost is about 15. A good indicator for the maturity level of compost is the NO_3-N/ Nmin ratio. As describe above, young compost contains mainly only NH_4 as mineralized N. Mature compost contains almost only NO_3 as mineralized N. Based on results of Swiss studies, we recommend the use of the NO_3-N/Nmin ratio as a reliable maturity parameter. These analyses should be done just before using the compost, and shortly after sampling. In the Netherlands, the common test is based on Oxygen Uptake Rate. An even more accurate method to measure heat generation by the compost is through the use of microcalorimeter.

5

MANURES AND COMPOSTS

5.1 INTRODUCTION

Manures and composts both contain plant nutrients in complex organic forms that release slowly into the soil after their decomposition. Their application results in the sustained supply of nutrients for longer duration besides improving soil physical properties. Manures and composts recycle the plant nutrients and they contribute back to the soil that was taken in the form of produce to consume by human being.

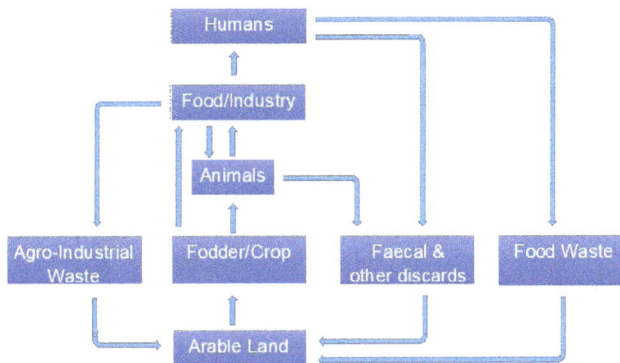

Fig. 1: Recycling of nutrients

5.2 MANURES

Manures are principally excreta of farm animals mixed with plant residues offered to the animals except for cakes, brown and green manures. However, many of the slaughterhouse wastes are also utilised as manures Major sources of manures are:

⊙ Livestock shed wastes viz. dung, urine and slurry from biogas plants etc.

⊙ Human habitation wastes viz.night soil, human urine, town refuse, sewage, sludge and sullage etc.

- Poultry droppings and litters etc.
- Slaughterhouse wastes viz. bone meal, meat meal, blood meal, horn and hoof meal, fish wastes etc.
- Agro-industrial by-products viz. oil cakes, bagasse and press mud, fruit and vegetable processing waste etc.
- Water hyacinth, weeds and tank silt etc.
- Brown and green manure crops and other manuring material

5.3 MAJOR CATEGORIES OF MANURES

Manures by the concentration of the nutrients may be grouped into two major categories bulky organic manures and concentrated organic manures based

5.3.1 Bulky Organic Manures

Bulky organic manures are poor in nutrients, so they are required in large quantities to meet the nutritional demand of the crops. Farmyard manure (FYM), and green-manure are the most important and widely used bulky organic manures. Advantages of using bulky organic manures:

- Bulky organic manures are required in large quantities, so they help in improving soil properties like structure, texture, porosity, water holding capacity etc.;
- They are a good source of micronutrients;
- They increase the availability of nutrients;
- They alter the microflora composition of the soil thus helps in controlling parasitic nematodes and fungi;

Commonly used bulky organic manures are:

- Farmyard manure
- Sheep and Goat Manure
- Swine Manure
- Poultry manure
- Green manure
- Green leaf manureand
- Brown manure

5.3.1.1 Farmyard manure

Farmyard manure (FYM) refers to the decomposed mixture of dung and urine of farm animals along with litter and leftover material from roughages or fodder fed to the cattle. On average well- decomposed farmyard manure contains 0.5 per cent N, 0.2 per cent P2O5 and .0.5 per cent K_2O. Urine has about 1.0 per cent nitrogen and 1.35 per cent potassium, which get wasted in the present method of preparing farmyard manure by the farmers. Nitrogen present in urine is generally in the form of urea is subjected to volatilisation losses. It is practically difficult to avoid losses altogether but can be reduced by following the improved method of preparing farmyard manure. Developing trenches to increase the absorption of nutrients available in the urine helps in increasing the nutritional status of the manure. All the available litter and refuse mixed with soil is filled and spread in the shed to absorb urine. Trenches of size 6.5 m to7.5 m length, 1.5 m to 2.0 m width and 1.0 m to 1.25 m deep are dug. The following morning, urine soaked refuse along with dung is collected and placed in the trench. A section of the trench from one side should be taken up for filling with thedaily collection. Whenthe section of the trench is filled up to a height of 45 cm to 60 cm above the ground then, the top of the heap is finished into a dome and plastered with cow dung and mud slurry. The process repeated for the second trench after the first trench is filled. The manure will be ready for use in about four to five months after plastering.

Fig. 5: Finished Farm Yard Manure (FYM)

5.3.1.2 Sheep and Goat Manure

The droppings of sheep and goats have higher nutrients than farmyard manure and compost. On an average, the manure has 3 per cent N, 1 per cent P2O5 and 2 per cent K2O. It is used to the field in two ways. The sweeping of sheep or goat sheds may be placed in pits for decomposition and can be applied later to the field. The nutrients available in the urine are wasted in this method. The second method is the penning of sheep, wherein sheep and goats are kept overnight in the field and urine and faecal matter added to the soil is incorporated to a shallow depth by working blade harrow or cultivator.

5.3.1.3 Swine Manure

Pig rearing is very common in the Northeastern region of India, and almost every rural household rears one or two pigs in their farm. The waste available from swine production can be converted to manure, and that can be an asset to a pork producer and agriculture when properly managed and utilised in a sustainable food production system. Swine manure contains all the essential plant nutrients that are used by plants. These include nitrogen (N), phosphorous (P), potassium (K), calcium (Ca), magnesium (Mg), sulfur (S), manganese (Mn), copper (Cu), zinc (Zn), chlorine (Cl), boron (B), iron (Fe), and molybdenum (Mo) (Chastain, 2003). The composition and efficiency of swine manure as a source of plant nutrients depends on several factors including the feed, supplements, medications, housing system, the method of manure collection, storage and handling. An adult pig produces about 6.0 kg of waste per day having dry matter content about 10 to 12 per cent. On dry matter basis swine manure contains 8.1–11.6 per cent nitrogen, 2.58–4.38 per cent phosphorus, 2.81–5.50 per cent potash, 3.12–6.33 per cent calcium, 0.93–2.66 per cent magnesium and 1.61–3.66 per cent sodium. Application of swine manure increases nitrogen, phosphorus, potash, calcium, magnesium and sodium availability in soil and enhances the yields of crops and vegetables. The efficient use of swine manure can be an agronomically and economically viable management practice for sustainable in temperate regions.

5.3.1.4 Poultry Manure

Among all animal manures, poultry manure's has long been recognised as most desirable manure because of its' high nitrogen and phosphorus content. It is mainly because solid and liquid excreta are excreted together resulting in no urine loss. Poultry excreta ferments very quickly. If left exposed, 50 per cent of its nitrogen is lost within 30 days. Poultry manure is used as a source of N, P andK butlitter also contains Ca, Mg, S and some other micronutrients. The average nutrient content of the poultry manure is 3.03 per cent nitrogen (N); 2.63 per cent phosphate (P_2O_5) and 1.4 per cent potash (K_2O).

5.3.1.5 Green manure

Green manure is a term commonly used to describe the group of plants or crops that are generally grown and then turned into the soil for restoring the soil fertility and increasing the organic matter content in the soil. Green manure crops usually belong to the leguminous family as the plants of this family can fix nitrogen in the soil with the help of symbiotic bacteria in their root nodules. Green manure crop can be grown in situ or may be brought from outside and incorporated in the soil. Green manuring crops can be cut and then tilled back into the soil or just left over the ground for degeneration. The most important green manure crops are Sunnhemp (Crotalaria juncea), Dhaincha (Sesbania bispinosa), Riverhemp (Sesbania rostrata), Clusterbeans (Cyamopsis tetragonoloba), Mung bean (Vigna radiate), Pillipesara (Vignatrilobata), and Cowpea (Vigna unguiculata).

Fig. 6: Green manuring with Dhaincha (Sesbania)

Nutrient contribution to the soil through green manuring depends significantly on type and stage of the crop. Biomass production and N accumulation of some common green manure crops are presented in table 1, and nutrient contents are presented in table 2

Table 1: Biomass production and N accumulation of green manure crops

Crop	Age(Days)	Dry matter(t/ha)	Naccumulated
Dhaincha	60	23.2	133
Sunnhemp	60	30.6	134
Cowpea	60	23.2	74
Pillipesara	60	25.0	102
Cluster bean	50	3.2	91
NewDhaincha (Sesbaniarostrate)	50	5.0	96

Table 2: Nutrient content and C: N ratio of major green manure crops

Greenmanurecrop	Nutrient content (%) on an airdrybasis		
	N	P_2O_5	K
Sunnhemp(*Crotolariajuncea*)	2.30	0.50	1.80
Dhaincha(*Sesbaniaaculeata*)	3.50	0.60	1.20
Sesbania(*Sesbaniaspeciosa*)	2.71	0.53	2.21

Advantages of Green Manuring

- Supplies good amount of nitrogen to the soil because most of the green manure crops belong to the leguminous family that has an association with nitrogen-fixing bacteria;

- The deep root system of green manure crops allows cycling of nutrients that have been leached to lower layers of soil;

- Green manuring enhances the soil aggregates stability and porosity, thus improves soil structure.
- Green manure increases water holding capacity by covering the soil and adding organic matter;
- Green manure also helps in reducing soil erosion as they increase the soil cover;
- Increased soil organic matter content favours the growth of microflora and macroflora and fauna;
- The release of organic acids during decomposition helps in solubilising stable phosphorus and make them available to the subsequent crop.
- Green manuring helps in the reclamation of alkaline soiland minimises root-knot nematodes population;
- They can serve as a lure crop to attract insect pests, thus helps in breaking the disease cycles;
- Green manuring helps in reducing the weed population as they compete for resources as well some green manuring crops also releases chemicals that are allelopathic to the weeds.
- Further, green manure crops are generally grown in the offseason, so it reduces weed proliferation and weed growth.

Disadvantages of Green Manuring

- Lack of interest of farmers;
- Green manuring involves cost and farmers don't see any direct financial return;
- Lesser development of breeding technologies for green
- Manuring crops;
- Poor availability of green manuring seeds;
- Some green manures crops may host diseases and pests that
- Mttack the commercial crop;
- The possibility of an adverse allelopathic effect of green manure residues on the commercial crop;
- The possibility of competition between green manure plants and the commercial crop due to inadequate management;
- Some green manuring crops have incompatible decomposition rates with the nutrient requirements of crops;
- Green manure crops can utilise moisture that may otherwise be conserved during fallow;

5.3.1.5 Green leaf manure

Application of green biomass including; green leaves and twigs of trees, shrubs and herbs collected from the vicinity are known as green leaf manuring. In the northeastern region, forest tree leaves are the main sources for green leaf manuring. Plants growing in wastelands, field bunds etc., are another source of green leaf manure. Some of the important plant species useful for green leaf manure and the nutrients available from them are summarised in table 3.

Table 3: Nutrient content of green leaf manure (Organic Farming, TNAU, Agritech Portal)

Plant	Scientific name	Nutrient content (%)on an air dry basis		
		N	P_2O_5	K
Gliricidia	Gliricidiasepium	2.76	0.28	4.60
Pongamia	Pongamia glabra	3.31	0.44	2.39
Neem	Azadirachta indica	2.83	0.28	0.35
Gulmohur	Delonixregia	2.76	0.46	0.50
Peltophorum	Peltophorumferrugenum	2.63	0.37	0.50
Weeds				
Parthenium	Partheniumhysterophorus	2.68	0.68	1.45
Waterhyacinth	Eichhornia crassipes	3.01	0.90	0.15
Trianthema	Trianthemaportulacastrum	2.64	0.43	1.30
Ipomoea	Ipomoea	2.01	0.33	0.40
Calotrophis	Calotropisgigantea	2.06	0.54	0.31
Cassia	Cassiafistula	1.60	0.24	1.20

5.3.1.5 Brown Manure

Brown manure involves growing of a pulse crop and knocking it down by spraying any herbicide to prevent weed seed set and maximise nitrogen fixation. Brown manuring is different from green manuring. In green manuring, the crops and weeds are killed by turning them to the soil, whereas in brown manuring crops are knocked down using a herbicide. Crops suitable for brown manuring includes all the green manuring crops like; Sunnhemp, Dhaincha, Mung, Cowpea, Lentil etc.

Figure 7: Sowing of sesbania with paddy for brown manuring

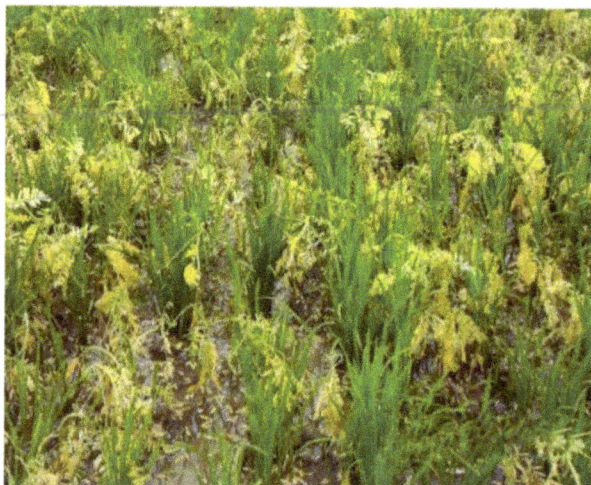

Figure 8: Knocked down plants of sesbania by spraying 2,4-D

Brown manuring technique can be well adopted in directly sown rice crop where Sesbania seeds are broadcasted @ 20 kg/ha, three days after rice sowing. The crop is allowed to grow for 30 days then it is knocked down by spraying 2,4-D ethyle easter. Sesbania shreds leave in 3-5 days. These leaves supply nitrogen about 35.00 kg/ha.

Table 4. Effect of brown manuring on soil organic carbon (C) and available postharvest nitrogen (N)

Year	Initial Ccontent of oil(%)	C content after harvest (%)	% increasein C	Initial soil available N content (kg/ha)	Soil available Ncontent after harvest (kg/ha)	% increase in soil available N
2014	0.54	0.69	0.15	283.0	320.2	13.1
2015	0.58	0.71	0.13	285.3	324.6	13.7
Mean	0.56	0.70	0.14	284.19	322.4	13.4

Advantages of brown manuring:

- Enhanced soil corban status;

- Savings in nitrogen fertiliser requirements as brown manure makes available organic nitrogen to soil;

- Provision of organic N, which acts as a slow release form of nitrogen for following crops;
- A manure crop fixes about 35 kilograms of nitrogen/ha;
- Helps in water conservation, due to increasing ground cover and weed control during the growing season;
- Other than nitrogen, available phosphorus and sulphur can also help in retaining and building soil carbon that improves soil resilience, water availability and productivity;
- Reduces the cost of cultivation
- After shedding the leaves, standing stems of degenerating crop acts as perch for predacious birds so helps in controlling insects.

Disadvantages of brown manuring
1. Prolonged use of herbicide may cause herbicide tolerance inthe crop;
2. Incurs additional cost as seeds of brown manure crop and herbicides.

5.3.2 Concentrated Organic Manures

Organic manure having higher nutrient content are termed as concentrated organic manure. In general organic manures containing nitrogen upto two per cent are included in the bulky category and those having more than two per cent nitrogen are treated as concentrated organic manure. Organic materials like oil cakes, fish meal, animal meal, poultry manures, slaughter house wastes etc. has comparatively higher contents of plant nutrients and are grouped under concentrated organic manures. These are also known as organic nitrogen fertiliser. Before the crops use their organic nitrogen, it is converted through bacterial action into readily usable ammonical nitrogen and nitrate nitrogen. These organic fertilisers are, therefore, relatively slow acting, but they supply available nitrogen for a longer period.

5.3.2.1 Oilseed Cakes

Oilseeds crops are cultivated for their oil content, and by-products obtained after oil extraction from the oilseeds are called oilseed cake. Oilseed cakes can be utilised either as feed for livestock and human or as a source of nutrients for the crops. Oilseed cakes are of two types, edible and non-edible. The cake obtained from edibleoil-bearing seeds and has potential to meet a part of the nutritional requirement of the livestock or human are called edible oil cake, and on other hands, those cakes, which has some antinutritional factors and toxic compounds and can't be used as feedstuffs are termed as non-edible cakes.

Fig. 9: Different types of edible and nonedible cakes

1. Edible oilseed cake: The cakes that can be utilised as feed stuffs for human or animals consumption are called edible cake. Oilseed cakes such as palm kernel cake, sesame cake, linseed cake and coconut cake contain 14–20 % of crude protein, where as soybean cake and groundnut cake contains 40–50 % of crude protein. These cakes are usedas feed supplements for human as well as animals; and

2. Non-edible oilseed cakes: The oilseed cakes, which has large doses of antinutritional factors and toxic compounds are not fit for feeding livestock or human consumption are called non- edible oilseed cakes; e.g., neem cake, jatropha cake, mahua cake karanja cake, castor cake, etc.

3. Both edible and non-edible oil cakes may be used as manures. However, edible oil cakes are offered to cattle as feed, and non-edible oil cakes are used as manures especially for horticultural crops. Oil cakes mineralise quickly, so they release nutrients in 7 to 10 day to the crops after application. Oilcakes need to be finely powdered before application for even distribution and quicker decomposition. The average nutrient availability of different oil-cakes is presented in the following table.

Table: The average nutrient content of oil cakes (Organic Farming, TNAU, Agritch Portal)

Oil-cakes	Nutrientcontent(%)		
	N	P_2O_5	K_2O
Non-edible oil-cakes			
Castor cake	4.3	1.8	1.3
Cottonseed cake (undecorticated)	3.9	1.8	1.6
Karanjcake	3.9	0.9	1.2
Mahuacake	2.5	0.8	1.2
Safflowercake(undecorticated)	4.9	1.4	1.2
Edible oil-cakes			
Coconut cake	3.0	1.9	1.8

Cottonseed cake (decorticated)	6.4	2.9	2.2
Groundnutcake	7.3	1.5	1.3
Linseed cake	4.9	1.4	1.3
Nigercake	4.7	1.8	1.3
Rapeseed cake	5.2	1.8	1.2
Safflowercake(decorticated)	7.9	2.2	1.9
Sesamecake	6.2	2.0	1.2

5.3.2.2 Other Concentrated Organic Manures

There are many slaughterhouse byproducts; they can be utilised as manure; e.g. blood meal, meat meal, horn and hoof meal, bone meal, fish meal etc. These meals are also used in the animal feed industry. The meals unsuitable for livestock consumption can be used as manure to supplement nutrients. They are a good source of nitrogen and minerals.

Fig. 10: Different (Blood, Bone, Fish & Meat) kinds of meals

The average nutrient content of animal-based concentrated organic manures is givenin the following table.

Table: The average nutrient content of animal-based concentrated organic manures (Organic Farming, TNAU, Agritch Portal).

Organic manures	Nutrient content(%)		
	N	P_2O_5	K_2O
Blood meal	10 - 12	1 - 2	1.0
Meatmeal	10.5	2.5	0.5
Fishmeal	4 - 10	3 - 9	0.3 - 1.5
HornandHoofmeal	13	-	-
Raw bone meal	3 - 4	20 - 25	-
Steamedbonemeal	1 - 2	25 - 30	-

5.4 COMPOSTS

Compost is a mixture of properly decomposed organic matter that can be used as a supplement for soil conditioning and supplying the nutritional requirements of the crops. It helps in better plants growth and enhances the yield of the crop. Crop residues, kitchen and farm wastes currently make up about 30 to 40 per cent of our daily trash, which we throw away. If these composted then at least it can meet the half of the nutritional requirements of the crop and keepingthese materials out of the landfills will help in cutting off the production of potent greenhouse gas, the methane.

Composting is the process of physical, chemical, and biological decomposition of large, bulky, coarse organic waste of plant or animal origin to a dark brown or black coloured, homogenous, brittle particulate. The matter obtained through this is called compost that can be used as a soil conditioner and source of nutrient to the plants. It naturally fortifies the soil and helps in enhancing the productivity of the soil.

Nutritive value of the compost depends upon the substrate used for composting. The compost prepared from farm waste like sugarcane trash, paddy straw, weeds and other plants and other farm waste has average nutrient contents as 0.5 per cent nitrogen, 0.15 per cent phosphate and 0.5 per cent potash.The compost prepared from town refuses like night soil, street sweepings and dustbin refuse etc. contain 1.4 per cent nitrogen (N), 1.00 per cent phosphate (P_2O5) and 1.4 per cent potash (K_2O).

Traditionally farm compost is made by placing farm wastes in trenches of variable sizes. Size of the trenches depends upon the quantity of the waste available for composting. Generally, 4.0 m to 5.0 m long, 1.5m to 2.0 m wide and 1.0 m to 1.5 m deep trenches are preferred. Available farm waste may be chopped to approximately 10 – 20 cm as chopping increases the surface area for microbial degradation. The chopped mass is placed in the trenches layer by layer. Each layeris moistened adequately by sprinkling cow dung slurry or water. Use of cow dung slurry adds additional nutritive value and also enhances decomposition rate. Trenches are filled up to the height of 0.5 m above the ground, and then it is left for decomposition. Sometimes water is also sprinkled over the decomposing mass for maintaining the moisture loss. In trenches improper or reduced oxygen supply restricts the decomposition process that lengthens the duration of composting.Compost becomes ready for application within five to eight months.

Composting is basically a biological decomposition of organic residues, however physical and chemical treatment is given to hasten the process. Depending upon composting methods, there are different kinds of compost and each has their advantages and disadvantages. These methods are as follows:

- ◉ Pit Manure/Anaerobic Compost
- ◉ Coimbatore method
- ◉ Indore method

- ⊙ Bangalore method

- ⊙ NADEP Composting

- ⊙ Vermicompost

- ⊙ Vermiwash

- ⊙ Biodynamic composting

- ⊙ Biodung composting,

- ⊙ Padegaon method,

- ⊙ Biogas-slurry method,

- ⊙ Azolla compost

- ⊙ In-vessel composting

- ⊙ Spent Mushroom Substrate (SMS)

- ⊙ Humanure

5.4.1 Pit Manure/Anaerobic Compost

Compost making in pits is the most common method practised in rural India. Length and width of the pit vary with the quantity of refuge produced in the farm and household, but usually, depth is kept about 0.75 to 1.0 m. This method had been prevalent primarily because it requires least investment and labour and secondly all kinds of degradable waste can be dumped in the pit for making compost.

Fig. 11: Pit composting

At the time of pit filling, aerobes are common micro-organisms present in the compost heap, and they need oxygen to grow, multiply and decompose the waste materials. But when there is insufficient oxygen supply in a pit, these aerobic micro-organisms can't survive, and they eventually die. After the death of aerobes, the decomposition process is taken over by micro-organisms called anaerobes, and they produce anaerobic compost. Nitrogen-rich materials are better for anaerobic composting, but almost any organic material can be processed including waste paper and cardboard, grass cuttings, left over food, animal slurries and manure, etc. The substrate with a lot of dry matter and high in corban is not preferred. In pit method decomposition is mostly anaerobic as composted mass is not turned regularly, which also leads to compaction. Usually, compost becomes ready for application in 6 to 8 months in this method but, quality of the compost is also very inferior. Presence of pathogens and weedseeds are the major disadvantage of this compost as they may not be killed, because anaerobic composting is a low temperature process. They may slowly become inactive due to a hostile environment and may not the potentially harmful.

5.4.2 Coimbatore Method

In this method, pits are prepared in the shaded area. Size of the pit depends upon the volume of organic waste available for composting, however, a pit of 3.5 × 1.8× 1.0m is preferred. A layer of farm wastes such as straw, vegetable refuse, weeds and leaves are spread to a thickness of 15-20 cm, then 5 cm thick layer wet cow dung is spread over it. Water is sprinkled over it to moisten the material. The process of filling the pit alternatively with farm waste and dung is repeated till the height of the material rises 60-75 cm above ground. After filling the pit, entire mass is covered with mud and plastered to facilitate anaerobic decomposition commences. In six to eight weeks matter partially decomposes and heap flattens. At this stage mud plaster is removed and the entire mass is turned, moistenedandleft for aerobic decomposition. The compost becomes ready to use in about four to six months.

5.4.3 Indore Method

Indore method was developed by Howard and Wardin (1931) in Indore (M.P) in India. Hence it is called the Indore method of composting. Length of the pit depends upon the waste available for composting; however, depth is kept as one meter, and width varies in between 1.5 to 2.0 m. The raw materials for making compost in this method includes plant residues, animal dung and urine, soil, wood ash and water. All organic wastes available on the farm; like crop residues, weeds, leaves and pruned part of the plants, and leftover fodders, are collected as a heap. Soft and succulent material are allowed to wilt to reduce moisture before stacking.

The pit is filled with 6-10 cm thick layer of organic wastes followedby 6-10 cm thick layer of urine soaked material along with dung and a thin layer of wood ash, soil and well decomposed old compost. Old compost acts as inoculum and hastens the decomposition process.These three sublayers together make a single layer and filled by 5-6 such layers fill the compost pit. At every 15 days interval, the composted material is turned to ensure proper decomposition. Water is mixed to moisten the material after each turning. Compost becomes ready for application in about four months.

5.4.4. Bangalore Method

C. N. Acharya primarily develops Bangalore method of composting in 1939 for safe composting of night soil and refuse. In this method, the refuse and night soil put in the pit. Length and width of the pit depend upon the quantity of material to be composted, and depth is kept about 0.9-1.2 m. The pit is filled alternately with night soil and refuse, and its level is raised 0.5 m above the ground. It is left uncovered for 15 days. During this period the material settles down and additional refuse and night soil is placed on the top then turned, moistened and plastered with wet mud. Filled and plastered pit left undisturbed for about 4-6 months. This system of composting provides aerobic conditions during initial days, and aerobic decomposition causes the rise in temperature for the first few days, which is retained for about 15 days, which helps in killing the pathogens and controlling the foul smell. After turning and sealing with mud anaerobic decomposition starts. The anaerobic decomposition is a slow process and takes 4-6 months to complete breakdown of organic matter.

5.4.5 NADEP Compost

The NADEP method of composting was developed by a farmer from Yavatmal district of Maharashtra, N D Pandharipande, who was popularly known as "Nadep kaka". In NADEP word NA stands for Narayan, DE for DEverao and P for Pandharipande. Being simple and ease in making compost, the method became very popularamong the farmers and termed as NADEP composting method using initial letters of his name. Nadep kaka was decorated with the RamabaiParkhe Award of Maratha Chamber of Commerce, Pune in 1983 for his contribution.

NADEP composting is aerobic decomposition of organic wastes placed in layers. The process involves filling of diverse compostable material in layers in specially designed pits. It helps in faster decomposition of organic waste as the design of the pit allows better aeration of composted waste. The quality of the compost depends basically on the type and ration of the substrate used for composting.

Fig. 12: NADEP Tank

The NADEP tank can be prepared as a permanent structure using brick, cement and sand or as a temporary structure using mud instead of cement and sand. The rectangular brick tank is constructed with maintaining enough space, i.e. honeycomb structure, for aeration between the bricks. Size of the tank prepared varies in length ×breadth× height, but 3 × 1.8 × 0.9 m is the preferred size for NADEP tank. However, if more material is available for composting, then length may be further extended, but it is advisable not to increase breadthbeyond 1.8 m. In making tank if mud has been used as mortar, then it is advisable that first two layers of brick should be done with cement to provide more structural stability and securing the structure from damage during filling and clearing the tank.The location of the tank is of utmost importance. Construction of the tank in the low lying area should be avoided. It can be placed at the high elevated area near some water source and preferably under the tree. The tank should be placed squarely in the direction of the wind. Old brick bats can be hammered, or a layer of bricks can be laid in the bottom of the tank to avoid seepage loss.

Materials suitable for composting include crop residues, dried husk, twigs, stalks, roots, leaves, dung, biocompost slurry, urine- soaked bedding material of livestock farm, well-filtered earth from irrigation channels, urine soaked earth form livestock farm etc. and water to moisten the organic wastes while filling them in layers.

While filling the tank care should be taken that it should be filled in one go and sealed within 24 hours. Before charging the tank with the materials, it is advisable to wet the inner walls and the tank bed with cow dung dissolved in water. After that, the first layer of plant waste of about six inches is filled. Above this, a mixture of 5-8 kg cow dung in about 80-100 litres of water is sprinkled. If biogas slurry is available then instead of cow dung, it can be used. The quantity of water utilised depends upon the season of the year and moisture percentage in plant waste being filled. If the moisture content is high in organic waste, water requirement will be less or vice-versa.

Fig. 13: Filled of NADEP Tank

Further, more water is required in summers than rainy or winter season. After the layer of cow dung, organic waste is covered with alayer of about 50-60 kg clean, filtered soil. Water is again sprinkled on it to maintain the proper moister content of the filled mass. After that, the tank continues to be filled with this series of layers in the same sequence. During the process of filling, it is advisable to put 2 to 3 bamboo logs vertically in the middle of the tank having one- meter distance from each other. These logs are removed generally after a month from filling the tank. The hole created after removing the logs helps in aeration and moistening the organic mass. Level of the filled mass is raised about one and a half feet above the rim of the tank. Usually, a tank can take 8-10 such series of layers. Once the filling is completed, then the tank is sealed with either mud or mixture of mud and cow dung. Care should be taken to avoid cracking of the plaster and if cracks appear the same may be plastered again with liquid cowdung slurry.

Fig. 14: Mud plastered NADEP Tank

With the commencement of microbial action decomposition of organic matter begins. This results in shrinking and material will shrink down below the tank rim is a period of 15 to 20 days. The tank should be opened and filled again in the same sequence of layers up to a height of 30 - 45 cm above the tank rim. Once again, the material should be covered with 6-8 cm thick layer of soil and sealed with liquid cowdung slurry. Level of filled mass is kept higher at the centre from the rim to avoid gathering of water in the middle either due to rains or during moistening. Accumulation of water may cause rotting instead of decomposing the mass.

Moister level of 15-20% is maintained to facilitate proper decomposition. Water may be sprayed through the holes on the tank sides and middle to maintain proper moister level. The entire tank can be covered with a thatched to prevent moister loss due to evaporation and the same time it also protects from rains. At no time compost be allowed to become dry, and cracks should not be allowed to develop and if the cracks develop same may be filled up with slurry. Grass that sprouts should be removed regularly.

Depending on the composting matter and way of preparation, the compost becomes ready in between 90 to 120 days. Status of compost can be checked from the side holes. When the composted mass turns to deep brown coloured crumbly particulate matter having a pleasant smell, then the tank may be opened as compost is ready for application. Compost may be removed for application. Each tank can be used for thrice in a year.

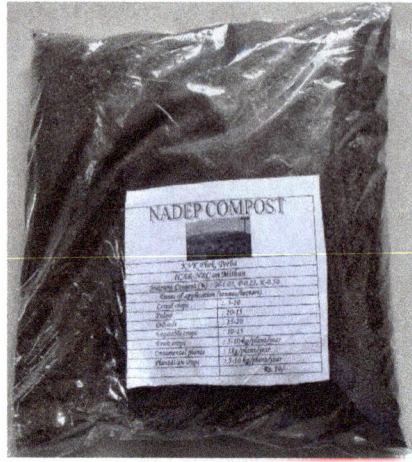

Fig. 15: Finished and packed NADEP compost

5.4.6 Vermicompost

Vermicomposting is a method of composting, where earthworms are used to digest and decompose the organic material. Earthworms, which generally lives is carbon-rich soils are used in making vermicompost. They eat organic biomass and excrete nutrient- rich castings, which is used as a fertiliser and soil conditioner. The digested organic matter excreted as worm casting is rich in humus and contains readily available plant nutrients. While passing from the earthworm's intestine, many of the pathogens and weed seedsalso get destroyed, and same time various enzymes and digestive juices released during the digestive process also come along with the cast. Further, the remains of worms after completing their life cycle also get mixed with the final product. These all enrich the compost and makes it more worthy to farm.

There are about 4,400 identified earthworm species, but only a handful of them are ideal for recycling organic waste. Earthworms can be classified into three ecological categories – anecics, epigeics and endogeics. The anecics and endogeics are known soil dwellers and may dramatically influence thesoil properties and processes at the ecosystem level, however, the functional role of epigeics is primarily that of litter transformers, like other litter invertebrates. Thus these epigeic earthworms are useful for vermicomposting.

Anecics (Gk-up from the earth) are dominant earthworms in many temperate region soils, (Lavelle 1983) are primarily vertically burrowing species. The burrows of these species may reach up to 3 m into the ground. They usually come to the soil surface only at night in search for decaying material like leaves.They pull their food down into their burrows and feed on it in the safety of their home. The feeding and casting habits of anecics may deeply influence soil characteristics up to >1m depth. Anecics are the largest worms on the earth and may grow several meters long. Lumbricusterrestris and Apporectodea longa are the typical examples of Anecic earthworms.

Epigeics (Gk-upon the earth) live virtually above the soil in the top 10 to 20 cm depth. They feed on decaying plant residues, dung and other animal waste.They are usually brown or red, which serves as camouflage and helps them to protect from their natural enemies. They consume, comminute and partially digest surface litter and rarely ingest soil particles. Epigeics feed purely on litter and generally have a short gut, so they greatly depend on a rapid response of gut microbes for fast digestion. Epigeic earthworm guts preferen-tially stimulate some microorganisms and reduce others that leads to a relative dominance of specific microorganisms that found in unin- gested soils. Examples of this group include Dendrobaenaoctahe dra, D. attemsi, D. rubidus, Eiseniellatetraedra, Eudriluseugeniae, Heliodrilusoculatus, Lumbricusrubellus, Perionyx excavatus, Pe rionyxarbicola and Eisenia fetida.

Fig. 16: Earthworm Eisenia fetida

Endogeics (Gk-within the earth) live in the upper layers of the ground and feed on the organic matter in the topsoil. They dig horizontal burrows when moving through the soil and rarely come to the surface. They are the most common earthworms (in biomass) in most tropical environments (Lavelle 1983). Endogeics are geophagous earthworms that feed on subsurface soil horizons. Their casts generally have more clay and organic matter than uningested soil. They help is releasing significant amounts of nutrients and NH4 in the soil. This group include Allolobophorachlorotica, Apporectodeacaliginosa, A. icterica, A. rosea, Murchieonamuldali, Pontoscolexcorethrurus and Lampitomauritii.

Among all the earthworms, the epigeics are the only ones that can be used commercially in vermicomposting. Lumbricusrubellus, Eudriluseugeniae and Eisenia fetida are the common worms, whichare being used commercially in vermicomposting. Among them, Eiseniafetida can be held in captivity. It feeds heavily on the rich organic matter like kitchen scraps, garden waste or compost and multiplies rapidly, these all factors make this worm an ideal choice for vermiculture.Almost all types of non-toxic organic matter, which are biologically degradable and decomposable can be used for vermicomposting,e.g. straws of cereals, pulses and oil crops, maize stalk, dry leaves, kitchen waste, pig manure, poultry droppings, rabbit manure, partially decomposed dung of cow, buffalo, mithun, etc. Composting materials should be screened for stone, metal, plastic, glass or any other non-decomposable materials. After cleaning the decomposable materials should be chopped into small pieces of 5-10 cm and to avoid unwanted organisms and foul smell materials are exposed to the sunshine for 1-2 days.

There is no shape, size or style specification for the vermicompost unit. It can be prepared in raised heaps above ground, in concrete tank or pits, in low-costlocal made bamboo structures, in precast cement rings, or in any such other containers, but covering with a shed over the structure is a must to avoid direct exposure to sunlight and rains.

Fig. 17: Vermicompost unit made of bricks

Fig. 18: Vermibag made up of polythene The procedure adopted for vermicomposting is as follows:

◉ Spread 20-30 cm thick layer of composting material (tree leaves, farm or kitchen refuge etc.) for aeration.

Fig. 19: Different types of substrates for vermicomposting

◉ Above that layer spread partially decomposed dung (20- 30cm thick) of about 1-2 months old (Use of fresh dung is not advisable as it starts microbial decomposition and raises the temperature of compostable material).

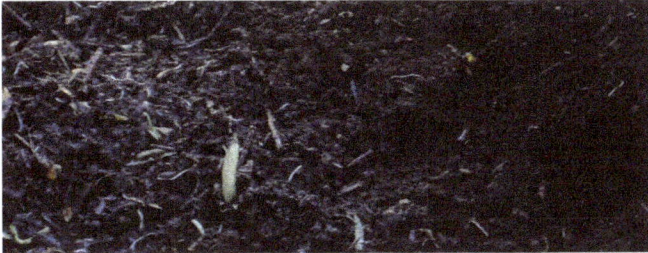

Fig. 20: Partially decomposed mithun dung

◉ Compostable material is sufficiently moistened and left for 15-20 days.

◉ Earthworms are introduced in it after 15-20 days at the rate of about 2-3 kg earthworms per ton of biomass or 1000 numbers of earthworms per sqm area.

Fig. 21: Inoculation of earthworms

◉ After introducing the earthworms another layer of feeding materials, i.e., kitchen waste, crop residues, cattle dung etc. are added.

◉ The whole mass is covered with gunny bags to prevent birds from picking the earthworms.

Fig. 22: Covering with gunny bag

- Maintaining moister up to 50-60% is advisable for the rapid growth of worms and faster decomposition. So, depending on the temperature and humidity, watering should be done to maintain proper moisture level.

- Feeding materials may be added weekly, and it should be turned fortnightly to facilitate proper aeration to the bin.

Fig. 23: Weekly addition of substrate and turning

- The whole biomass is converted to vermicompost in about 2-3 months.

- When the mixture becomes brownish or blackish brown, and there does not remain any odour of dung, then the vermicompost is ready.

- Hence, adding feeding material and watering should be stopped, which will result in the migration of worms to the lower layers.

- Remove the compost, separate the worms and keep the material in the shade for drying. The final product Vermicompost is granular and looks like tea leaves.

Fig. 24: Sieving of finished vermicompost

Fig. 25: Finished product

⦿ Sieve out the compost and fill in the plastic bags for use.

Fig. 26: Finished and packed Vermicompost

⦿ After sieving, the remained material has a good number of cocoons, so it can be used as inoculum for next pit or remained material may be moistened and kept separately in the shed for another 20-25 days. Each of the cocoons has 1-20 eggs, and that may develop into young earthworms in 2-3 weeks. These young ones can be separated and may be used as inoculum. Cocoons of earthworms are required to produce new young earthworms that replace the aged one. These cocoons must be saved to multiply the earthworms. Hence, the utilisation of suitable mesh size sieve is essential.

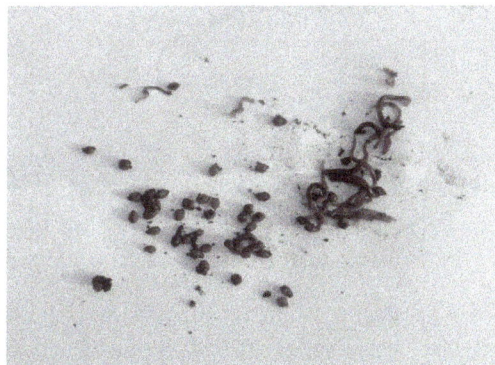

Fig. 27: Cocoons and vermicules

a. 3 mm mesh size sieve is better as nil or fewer numbers of cocoons can pass from the sieve during sieving (Murali, 2011).

b. The number of worms shall also be increased by 20-30 times from the number introduced earlier. The same may be collected and released in other prepared bins.

c. Ants, termites, flatworm, centipedes, rats, pigs, birds, etc. are important natural enemies of vermiculture, so preventive measures must be taken to avoid the loss.

d. Mixing of neem cakes @ 30 g/ 1kg food while filling the beds helps in keeping away the natural enemies.

e. Small furrows can be made, and water should be filled to avoid the loss from ants, termites and centipedes.

f. Covering with gunny bags keeps the birds away from them.

Nutrient Content of Vermicompost (%): Though availability of nutrients greatly depends upon the kind of substrate used for making vermicompost, but the average content may as follows:

7. Nitrogen (N): 1.5-3.0

8. Phosphate (P2O5): 1.5-2.5

9. Potash (K2O): 1.5-2.0

5.4.7 Vermiwash

Vermiwash is a liquid plant growth regulator, which contains a high amount of enzymes, vitamins and hormones like auxins, gibberellins etc. along with macro and micronutrient used as a foliar spray.

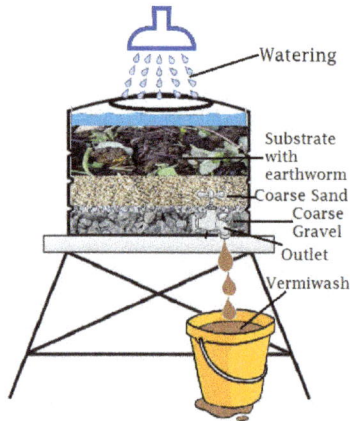

Fig. 28: Diagramatic representation of a vermiwash unit

A container (concrete/plastic) with a small hole at the base is used for making vermiwash. A layer of gravel/broken pieces of bricks is placed at the bottom of the container to the height of 10-15cm above which another layer of coarse sand (10cm) is placed. A layer of farm wastes and cowdung in the ratio of 1:1 is added to the container,which is followed by the release of earthworm species like Eisenia foetida, Eudriluseuginiae, Perionyx excavatus etc. Above which dry straw is to be added, and watering is done. After about ten days vermiwash starts forming in the container and that can be collected through the outlet.Vermiwashcan be used fresh or may be stored for a few days. It can be diluted with water and sprayed to the plants or may be directly drenched to the plant.

5.4.8 Biodynamic Composting

Biodynamics, derived from two Greek words, bios (life) and dynamos (energy). In this method of farming, farmers treat the farm as a living system, where plants interact with the cosmic energy, environment andsoilto produce food that nourishes, vitalises and helps to develop dynamic and strong societies.

Biodynamic composting is based on esoteric concepts drawn from the ideas of Rudolf Steiner developed in 1924. It emphasises on the holistic understanding of the interrelationships between soil fertility, plant growth, and livestock care. In this method fermented herbal and mineral preparations are used as compost additives and field sprays. Biodynamic preparations described by Steiner are used to make biodynamic compost. The biodynamic (BD) preparations are numbered from 500 to 508. The BD 500 and 501 are the field preparation, whereas BD 502 to 507 are compost preparations. However, BD preparation 508 had antifungal property and used as preventive and curative measures against the fungal problem.

The BD 500 preparation (cow horn manure) is made from cow manureby filling the horn of a cattle with cow dung manure and burying it 40 to 50 cm below the surface in the ground in the autumn. It is leftfor decomposing for six months and recovered after winter. It is used as a soil spray to stimulate root growth and humus formation. The BD 501 preparation may be prepared by stuffing a horn of thecow with crushed and powdered quartz and burring the same into the ground in spring. The buried horn is taken out in autumn. One tablespoon of this quartz powder can be dissolved in100 litres of water and may be applied as a foliar spray to stimulate and regulate plant growth.

Compost preparations BD 502 to 507 are used for preparing compost. Various kinds of herbs, which are frequently used in medicinal remedies employed to make these preparations. Details are as follows:

502: Yarrow (Achillea millefolium) flowers are dried and stuffed into urinary bladders of the stag (Cervus elaphus) and placed in the sun during summer. The stuffed urinary bladder is kept in the earthen pot and buried in earth during September and retrieved in March. It helps plants for better absorption of trace elements.

503: Chamomile (Matricaria chamomilla) flowers are dried and stuffed into small intestines from cattle. Filled sausagesare stacked in bundle and place in earthen pot, which is buried in the humus-rich earth in October and retrieved in February/March. This preparation helps in stabilising Nitrogen (N) within the compost and stimulates plant growth.

504: Stinging nettle (Urtica parviflora) plants in full bloom are stuffed together into terracotta pipes or earthen pots and bury the same underground surrounded by peat for a year in September. The preparation helps in stimulating soil health and enhances the nutrient absorption capacity of the plant.

505: Oak (Quercus glauca) bark is crushed and placed inside the brain cavity of the skull of a domesticated animal. Bury the skull in a watery environment with weeds and plant mulch or peat in September. It is retrieved in March and on retrieval there will be foul smell and fungal growth, which goes off on drying in dark anddry place. It helps in combating harmful plant diseases.

506: Dandelion (Taraxacum officinale) flowers are wrapped into the mesentery of a cattle and tied with a jute thread. The wrapped parcel is placed into an earthen pot filled with the properly mixed soil and compost. The pot is buried in earth during September and retrieved in March. This preparation helps in simulating relation between Si and K.

507: Valerian (Valeriana officinalis) flowers are ground with a mortar and pestle. The paste obtained is mixed with water in 1:4 ratio and stored in a cool place. This preparation helps in better utilisation of Phosphorus in the soil.

Finally, there is BD preparation 508 used as a foliar spray to suppress fungal diseases in plants. It is prepared from the silica-rich horsetail plant (Equisetum arvense) or Casuarina.1 kg dried Horsetall herbis boiled in 10 litres of water for 2 hours and kept for two days then the decoctum obtained is filtered and stored for future application. 50 ml of decoctum/tincture is diluted in 10 litres of water and sprayed onto the soil or over the plants during the early growing stage. It is effective against mild fungus and sooty mould problems.

Composting Process

The biodynamic compost is prepared as a surface heap rather than in a traditional pit. Elevated, flat, dry and away from tree shade site is selected for making the heap. A rectangular piece of land as per the quantity of compost required is marked and a set of logs or PVC pipes placed in the middle to facilitate proper aeration of the pile. A layer of 15-20 cm thick, dried and green biomass is stacked alternatively to make the pile for composting. After 2-3 layers soil from the fertile plot and cow dung slurry may also be added, along with it crushed rock phosphate and slaked lime is added in the middle layer to hasten the decomposition process and to supplement the mineral content. Stacking of the pile is done for about 1.00-1.25 meters high. After the compost pile isdone, BD preparations 502–506 are tactically placed 5–7 feet inside the pile, in holes poked about 30-40 cm deep. Preparation No. 507, or say liquid valerian, is applied to the outside layer of the compost pile by spraying or hand watering.

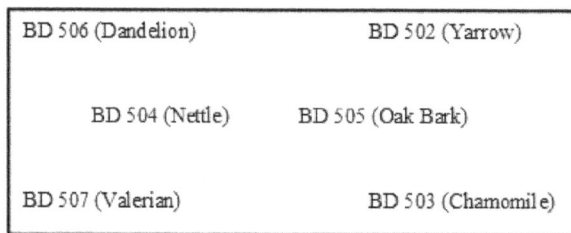

BD 506 (Dandelion)	BD 502 (Yarrow)
BD 504 (Nettle)	BD 505 (Oak Bark)
BD 507 (Valerian)	BD 503 (Chamomile)

Fig. 29: Placing of BD preparation in the pile

Once stacking is completed and the pile is charged with BD preparations, it is sealed with the paste made from clay and cow dung. Pile is watered regularly through the holes made for aeration and watering. If any crack appears in the plaster same is sealed to avoid rotting or drying. Piled material decomposes in about 2-3 months. At the end of this time, the compost is tested for its readiness. Readied compost smells like forest soil, which indicates that the degradation is almost complete and that the compost is ready to use. Alternatively while preparing NADEP compost BD preparations, 502–507 can also be used to get biodynamic compost.

5.4.9. Biodung Composting

Biodung composting is the combination of partially aerobic and partially anaerobic decomposition. This method was developed at the Centre of Science for Villages, Datapur, Wardha by Priti Joshi in 1996. In this method, weeds were used in the preparing the biodung compost. In this method, a systematic heap of 1 x 0.5 x 1 m is prepared by stacking the biomass in definite layers. Cow dung solution is sprinkled over each layer of the biomass. 5-6 layers of biomass are place done over another to make a heap. The heap then covered with black polythene sheet for retaining optimum temperature level (40 to 55 °C for 15 days). The temperatures may reach as high as 600C to 650C as decomposition starts. At this temperature, most of the parasites and pests and viable seeds of weeds get destroyed. As temperature goes down in about 30 days, this composted material can be used for vermicomposting. Biodung composting is generally applied as a prerequisite to vermicomposting.

5.4.10. Padegaon Method

The padegaon method is recommended for composting of hard to digest substrates like sugarcane trash and cotton stubbles. These materials are shredded into small pieces of about 30 cm and flattened to make a 30 cm thick and about 2 meter wide layer above the ground. Length of the layer depends upon the quantity of raw material available. The layers are drenched with the slurry prepared by mixing of wood ash, cow-dung and soil from a fertile field. Four to five such layers are added to the pile. Finally a heap of about

1.5 m high in raised and kept for decay. As the substrates used are very resistant to decay, the heap is turned each month and after turning it is trampled again. Water is added in sufficient quantity to keep the material moist. The material decays in 5-6 months and becomes ready for use. The padegaon compost compares very well in composition with farmyard manure.

5.4.11. Bio-Slurry Method

Biogas plants anaerobically digest cow dung, farm waste and other biodegradable materials to produce precious amounts of combustible gas, known as 'biogas'. At the household level, this biogas can be effectively used in gas stoves for cooking and lamps for lighting. Biogas produced in large digesters can be used for running engines, motors or filling biogas cylinders. The residue emanating from the biogas plant is known as 'bio-slurry'.

Fig. 30: Bio-Slurry emanating from biogas unit

It can be easily collected and used as a potent organic fertiliser for crops and aquaculture. In India biogas has replaced dried dung cakes for cooking and the bio-slurry has a distinct added value as it can still be used as fertiliser.

Bio-slurry is rich in nitrogen (N), phosphorous (P), potassium (K), and several micronutrients. The content of these plant nutrients varies with the type and nature of feed stocks used for the production of biogas. The digested slurry discharged from the biogas plants normally contains 92–94 per cent moisture whereas in the case of solid-state biogas plants, the moisture content varies between 88–90 per cent (Hazarika et al. 2015). The average nutrient and organic matter content of the bio-slurry is given in Table 1.

Table 1: Average nutrient content of bio-slurry

S.No.	NameofNutrient	AverageContent
1	N	1.8%
2	P_2O_5	1.0%
3	K_2O	0.90%
4	Mn	188 ppm
5	Zn	144 ppm
6	Fe	3,550 ppm
7	Cu	28 ppm
8	C/N ratio	10–15
9	Organic matter	65%

5.4.12 Azolla Compost

Soil health directly influences the growth and production of the crops. The health of the soil depends on anoptimum combination of organic and inorganic components of the soil. Regular use of chemical fertilisers adversely affects the soil biota and destroys them. There are many useful soil micro-organisms present in the soil, which can help plants to get nutrients. In this context in addition to all biofertilizers agents, Azolla also plays an important role. It acts as a nitrogen-fixing factory in standing crop.

Azolla, a small free-floating water fern, is available naturally on moist soils, the water surface of flooded rice fields, small ponds and canals. Its size is about 1.5-3.0 cm in length and 1.0-2.0 cm in breadth. Roots of Azollae manates from growing branches and remains suspended in water. It has some exciting attributes to be used as an aquatic green manure plant as well as animal feed

Fig. 31: Azolla (A. caroliniana)

Azollahas a symbiotic relationship with an N-fixingblue-green alga (Anabaena azollae) thus can fix atmospheric nitrogen into the soil and has the potential of providing 30-60 kg N/ha under normal field conditions. It can be grown on any closed body of water. Azolla being palatable and high in protein and minerals can be used as a feed to chickens, pigs, goats, ducks and cows.

Worldwide seven different species of Azolla have been identified. Out of them, A. pinnata is widely available throughout the Asian subcontinent and can be easily collected from ponds and ditches. Besides this, some better species like A. caroliniana can also be grown successfully particularly in NE India.

Conditions for Azolla growth: Water depth of 10-15 cm is necessary for multiplication of Azolla. Slightly acidic soil having pH of 5.2 to5.8 is good for the growth of Azolla. It prefers partial shade and can grow well at a temperature of 15-300C. All these conditions are available in rice fields, therefore as double cropping of Azolla with rice may be the most successful. Naturally, it can be seen growing in the rice field in NE region of India as soils are acidic.

Cultivation practices of Azolla caroliniana: Cultivation of A. caroliniana is detailed as follows:

- Dig out a pit of the size 2×1×0.20 meter and level the pit.
- Place a lining of good quality polythene to restrict percolation of water.
- Raise the bund all around the pit and maintain the water depth of about 10-20 cm
- Add 10 gm rock phosphate plus 200gm of dried powdered cowdung in the water

- ⊙ Inoculate 300-400 gm of fresh A. caroliniana in the pit
- ⊙ Leave it for multiplication for 15 – 20 days until a thick mat of Azollais formed.
- ⊙ By this time there will be an increase in fresh biomass to the tune of 8 -10 times.

Fig. 32:Azolla cultivation in LDPE/Cemented tanks

Composting of Azolla: Fresh Azolla can be collected from fields and ponds on the bunds and left for a day to drain the water. After draining the water, Azolla is stacked in a shaded area for 15-20 days. Since Azolla has an excellent carbon-nitrogen ratio, so it decomposes rapidly and converts into high-value compost within this period. On maturity Azolla compost has moisture about 20-25 per cent, Nitrogen (N) 2.60-3.50 per cent, Phosphate (P2O5) 0.60-0.70 per cent and Potash (K_2O) 2.80-4.00 per cent.

Fig. 33: Packed Azolla compost

5.4.13 In-vessel Composting (IVC)

In-vessel Composting is a way of disposing of organic refuse in large batches. It generally defines a group of methods that confine the composting materials within a building, container, or vessels. In-vessels take up less land, process organic waste faster and can handle a much more extensive array of substrates. This technique is generally used for handling waste from food, meat, and fish processing industries, municipal sewage biosolids, field and garden waste etc. It allows large-scale organic waste processing to a safe and stable product suitable for reclamation as a soil amendment.

IVC ensures that the composting takes place in an enclosed environment, with proper temperature, moisture, and airflow control and monitoring. There are many different IVC systems, but they can be broadly categorised into six types: containers, silos, agitated bays, tunnels, rotating drums and enclosed halls. These may be comprised of metal or plastic tanks or concrete bunkers inwhich air flow, moister and temperature can be controlled, using the principles of a "bioreactor". Generally, the air circulation is maintained through in via buried tubes that allow fresh air to be injected under pressure, with the exhaust being extracted through a biofilter, with temperature and moisture conditions monitored using probes in the mass to allow maintenance of optimum aerobic decomposition conditions. Decomposition occurs at a faster rate than the conventional composting processes, and the compost becomes readywithin a few weeks or month.

IVC system can also be integrated with anaerobic digestion. In this approach batches of organic material are first subjected to anaerobic digestion in the container/vessels and then later in the same bioreactor the process is switched to composting through the use of forced aeration.

5.4.14 Spent Mushroom Substrate (SMS)

Mushrooms the fleshy, spore-bearing fruiting body of the fungus are well known for their delicacy, nutritional and medicinal values, but the substrate left after the harvesting of the crop better known as 'Spent Mushroom Substrate' also have great importance. The substrate for mushroom growing is prepared by controlled fermentation from agricultural, poultry and industrial wastes. Primarily it is used for mushroom cultivation, and the spent substrate obtained after crop harvest possesses all essential attributes of organic manure and if this is not appropriately handled may create various environmental problems and become a nuisance.

The re-composted spent mushroom substrate has been found to be an excellent growing medium for the majority of the vegetables and the crops and has shown multifaceted utilities in improving the yield and quality of the crop, and management of the diseases, which is encouraging for the mushroom industry. The other utilities of the spent mushroom substrate, like in vermicomposting, bioremediationand as organic-mineral fertiliser are boon to the country's farming system.

Bioremediation of contaminated soil: SMS adsorbs the organic and inorganic pollutants from the soil. It also harbours diverse category of microbes that have capabilities to break down organic xenobiotic compounds biologically. The actinomycetes (*Streptomyces* spp. and *Thermomonospora* spp.) present in SPS also have strong pollutants catabolising capabilities that help in reducing pollutants level in the contaminated soil after incubation with SMS.

Nutritional composition of SMS: Nutrient status of SMS changes with withering. It contains Nitrogen (N) 1.9%, Phosphate (P) 0.4% and Potash (K) 2.4% before weathering, however after weathering for 8-16 months status changes to Nitrogen (N) 1.9%, Phosphate (P) 0.6% and Potash (K) 1.0 %.

Re-composting Methods: SMS being rich in organic matter and nutrients help in improving the soil carbon, neutralioco acidic soils and support plant growth. It not only improves the fertility of the soil but also enhances water holding capacity, porosity and texture on applying as manure. Re-composting of SMS is done to improve its physico-chemical propertiesfor using it as manure. Following three methods are generally employed for re-composting SMS.

Natural: In natural weathering process SMS is stored in the pits of about 1.25×1.25×1.25 meters and left for anatural weathering. It takes about 1 to 2 years for complete weathering.

Aerobic: A pits with the perforated base of about 1.25×1.25×1.25 meters is prepared. Generally, waste wooden logs are used in the base to achieve perforation. The perforated bottom of the pit is connected perpendicularly with 5 cm diameter hollow plastic pipes placed at a distance of 30 to 40 cm. Holes of about 2 cm diameter at a distance of 15 cm each in a circular fashion are also made on thepipe inserted perpendicularly for facilitating better aeration. The pit is filled with SMS up to the brim and left for decomposition.

Anaerobic: In this method, we fill the pit similarly as in natural method, but on the top, the pit is covered with 20-30 cm thick layer of normal soil and left as such for weathering.

5.4.15 Humanure

"Humanure" is a buzzword designated for human excrement (faeces and urine), which is recycled via composting for agricultural or other purposes. The term was first used in 1994 by Joseph Jenkins that in his book "The Humanure Handbook" where he advocates the use of humanure as a soil amendment.

Humanure is notthe "sewer sludge" or "biosolids" or "sewage compost" obtained from the sewage treatment plant as this may include various pollutant released from industries or other sources into the sewage line, instead, it is the decomposed product of faeces, urine, toilet paper and some additional carbon sources such as sawdust. Humanure also differs from night soil, which is raw human waste spread on crops. Night soil, while aiding nutrients present in faecal matter to the soil, it also contaminates the soil by spreading many human pathogens present there.

A humanure system like compost toilet works on zero energy as it does not require water or electricity. It does not smell if appropriately managed. Human excrement from compost toilet is collected and added to a hot compost heap together with sawdust and straw or other carbon-rich materials. The heat generated during the decomposition helps in destroying the pathogens. The waste is processed in situ in a composting toilet. Composting is the best way to dispose off the faeces and urine, the nutrients contained in them are returned to thesoil. It helps in nourishing the soil and preventing soil degradation. Human faecal matter and urine have high percentages of nitrogen, phosphorus, potassium, carbon, and calcium apart from micro- nutrients.

Humanure aids in the conservation of freshwater by avoiding the use of potable water required by the typical flush toilet. It further helps in preventing the pollution of groundwater as the faecal matter decomposes before entering into the system. There should be no groundwater contamination from leachate if appropriately managed. This technology is eco-friendly, reduces energy consumption thus checks greenhouse gas emissions associated with the transportation and processing of water and wastewater.

Use of humanure is safe in the crop if it is handled appropriately and composted adequately. Because at the time of decomposition, temperature increase and sufficient heat are generated that destroys the harmful pathogens or enough time must elapse since the fresh material is added so, the biological activity should kill any pathogens present there. During the process humanure making pathogenic organisms get killed by the extreme heat of the composting, i.e. above 45 °C for weeks and the extended amount of time (1 to 2 years) that it is allowed to decompose. Complete pathogen destruction is guaranteed by arriving at a temperature of 62 °C (144 °F) for one hour, 50 °C (122 °F) for one day, 46 °C (115 °F) for one week or 43 °C (109 °F) for one month. A curing stage is needed to allow a second mesophilic phase to reduce the potential phytotoxins to make it safe for crops. Therefore, it is advisable to use the humanure after elapsing one year time.

Choosing a Composting Method

Choosing the method that best suits for composting depend upon various factors. Before selecting the method of composting following questions must be addressed.

a. Who are you?
b. Why do you want to make compost?
c. Where are you located?
d. Available technological options
e. Kind of substrate available

Who is going to make compost is an important question because one method applicable with individual small farmer may not be suitable to commercial composter? Accordingly, a small farmer may be producing for his farm use; however, large/commercial composter may be doing so for sale of compost. Where you are located is also an important factor from environmental impact and potential nuisance viewpoints. For example, composters located in rural areas need to be concerned about environmental factor and assure that composting activity shouldn't adversely affect the surface and groundwater aquifers, however, they usually may be less concerned about neighbour complaints related to odours, noise and dust so they may go for low technology composting. However, composting sites located in rural and urban areas have to utilise high technology options to minimise nuisance.

6

ENRICHMENT OF ORGANIC COMPOST

6.1 INTRODUCTION

Enrichment of the compost is a way of fortifying naturally produced compost. It is an excellent way of enhancing the quality of compost. Most of the Indian soil is deficient in Phosphorus, and continuous cropping further drains it. Yearly removal of phosphorus from soil is more than the addition. Similarly, other nutrients are also drained from the soil. If the traditional technology of composting is improved in the term of its nutritional status, then it can help significantly in checking the nutritional depletion.

Fig. 1: Enrichment of compost/manure

6.2 COMMON ADDITIVES AND METHODS TO PRODUCE ENRICHED COMPOST

6.2.1 Clay Soil

Addition of 10% clay soil in the compost pile helps in reducing nitrogen losses and makes the end-product more stable. Clay acts as a "colloidal trap," and help in retaining nitrogen. Microbes present into the pile converts this nitrogen from a gaseous form to a usable form. Adding thin and repetitive layers of clay soil works better.

6.2.2 Rock Phosphate

Rock phosphate can increase the availability of usable phosphorus to crops and help in reducing gvolatilisation of ammonia. In India, about 160 million tons of rock phosphate deposits are available, but they are of low-grade and contains less than 20% P2O5, which are considered unsuitable for manufacturing commercial phosphatic fertilisers. However, this low-grade indigenously available rock phosphate works well in acid soils, however, modifications required to use in neutral and alkaline soils.

6.2.3

The Indian Institute of Soil Science (IISS), Bhopal has developed technology to utilise the rock phosphate in combination with phosphate solubilising bacteria (Aspergillus awamori, Pseudomonas straita, and Bacillus megatherium), pyrite and bio-solids. The phospho-compost/N-enriched phospho-compost technology enriches the manurial value of the compost compared to FYM and other ordinary compost.

6.2.4 Material Required

1900 kg organic waste, 200 kg cow dung (on dry basis) and 250 kg of rock phosphate (18% P2O5)are used to produce one ton of phospho-compost through this method. The final product contains about 2—3% phosphate, and it becomes ready for application in about 90-100 days.

6.2.5 Methods

Base of the heap is prepared from hard and woody materials like woody sticks, bamboos etc. Size of the base depends upon the quantity of the substrate available for composting. Generally, a base of 3×3×0.15 m is preferred. Above the base, 30-40 cm thick layer of bio-solids is placed. Over the bio-solid layer slurry prepared by mixing cow dung, rock phosphate and microbes are sprinkled. Another layer of the substrate is added and moisten with slurry. Stacking of the material continues with the alternate layer of crop residue and slurry until the heap is 1.5 m high. Area of each layer is reduced to taper the heap. Water is added to the heap to maintain moisture about 60 to 70% to facilitate proper bacterial activity and decomposition. Heap is covered with mud or polythene. The material is turned after 30 & 45 days. Add water at each turning to maintain the moisture content about 60-70%.The compost becomes ready for field application within 90-100 days period.

Table:1. Nutrient composition of manure and phosphocompost

Manure	Total N (%)	Total P (%)	C: N ratio
FYM	0.5-0.8	0.32-0.55	22.0-25.0
Compost	0.6-0.8	0.55-0.60	22.0-25.0
Phospho-compost	1.2-1.4	2.00-3.50	17.0-18.0

6.2.5.1 Enriching compost with waste-mica

Along with rock phosphatewaste mica generated during cleaning of raw mica can be usedto enrich the compost as it is a good source of potash and other micronutrients. The

Top dressing with a 1/8 to 1/2 inch layer of well decomposed compost isadvocated at least twice ayear in the gardens. Care should be taken to continue with the routine watering of the lawn asexcessivelydry condition prevents decomposition of compost (Bordoloi et al., 2015).

6.4.3 Placing manure in pits

Compost may be applied in pits during planting of crops. This method is found to be effective for vegetablecrops like ginger, turmeric, potato tubers, cucurbits, seedlings of vegetables such as cabbage, cauliflower, knolkholetc.

6.4.4 Compost as a potting mixture

Compost is mixed with soil as a potting mixture for application is flowering pot and small gardens. Generally, compost, sand and garden soil ratio is maintained as 1:1:1. Transferring seedlingo along with pot mixture to the field helpavoid stress to the seedlings besides other benefits.

6.4.5 Compost as mulch

Various kinds of mulches are used to conservemoisture, improve water balance and soil structure, control of weeds, minimize soil temperature fluctuations, increase soil fertility and reduce water erosion of soil. Three to four inches thick layer offinished or unfinished compost can be utilized as mulching material.

6.4.6 Use of Compost Tea

Water extract of compost is termed as compost Tea. It is rich in soluble micronutrients, humic acids and growth promoting substances. This tea can be sprayed to crops as a growth booster. Spraying compost tea is an excellent way to nourishindoor or outdoor potted plants. It can be prepared by placing a clothbag filled with mature compost in a bucket of water for an hour andsubsequently harvesting the liquid as compost tea. The remaining content of the bag can be utilized as a soil amendment, whereas compost tea for spraying.

6.5 CONCERNS WITH COMPOST AND MANURE

Compost helps in improving soil structure and nutritional status but sometimes it may also source in spreading of disease and weeds. Therefore it is always advisable to be precautious in producing and using the compost. Properly cured and produced compost is considered to be free for any harmful organism and weed seeds as during composting temperature of the composted mass rise as high as 650C, and that kills all the disease-causing microbes and weed seeds. However, during the process, it is possible that certain organism and weed seeds may escape due to faulty composting.

Further, one should not eat crops grown in areas where manure has been applied recently. It takes at least three months to neutralise the ill effect. Generally, pathogens present in manure, such as Salmonella, Listeria and E.coli, may contaminate the food. The risk of pathogens contaminating food is highest for root crops and leafy vegetables. The edible portions of these vegetables come in contact with contaminated soil thus passes the pathogens. Careful washing or peeling eliminates the risks, and cooking also effectively kills the pathogens. Also, raw or poorly decomposed manure should not be applied under the fruit trees because the fruit may fall there and get contaminated as it comes into contact with the manure.

Sometimes, poultry, rabbit and pigs are fed organoarsenicals to control coccidiosis and growth promotion, but the compost prepared form the faecal matter obtained from such farms may contain Arsenic (As). Repeated use of such compost for the extended period may result in arsenic buildup in the soil. Such soils become sick and may lead to uptake of arsenic by plants. It may also leach to groundwater, thus polluting the water and find the way in the human food chain.

7

VERMIWASH: A POTENTIAL TOOL FOR SUSTAINABLE AGRICULTURE

7.1. INTRODUCTION

In India, the first green revolution enormously enhanced the crop production, but on the other hand, massive Application of chemical fertilizers over a period has resulted in poor soil health, reduction on agricultural produces, and increases in incidences of insect pest and disease and environmental pollution , and long term use of various agrochemicals like fertilizers, plant growth promoters, pesticides and improved seed varieties, adversely affected ecosystems like soil, water, and food contamination and gene pool of wild seeds. The second green revolution started with organic farming. But for feeding the vast increasing population of our nation, immediately we cannot completely shift towards organic farming. Keeping this view in mind we have to go for sustainable crop production. The aim of sustainable agriculture is to fulfill our present needs (food, shelter and clothes) without compromising the ability of future generations to meet their own needs. Therefore, three main objectives of sustainable agriculture is a healthy environment, economic profitability, and social and economic equity, and for achieving these objectives application of vermiwash can play a important role in ensuring a sustainable agricultural

During first green revolution, excess use of agrochemicals adversely affects natural resources and ecosystem. Therefore, we must have to go for sustainable agriculture to conserve and protect our resources and ecosystem. For this vermiwash is emerging as important potential tool. Vermiwash is a brown coloured, odourless, liquid biofertilizer, which is collected after passes via column of worm culture. It is considered as storehouse of nutrients and microorganisms, used as foliar spray for crops. Vermiwash contains mucus, excretory products of worms and various concentration of macro, micro and beneficial nutrients along with beneficial microorganism, growth hormones, Vitamins, enzymes and amino acids, therefore is a good source for plant nutrition in sustainable agriculture.

system. Vermin wash is a liquid organic fertilizer obtained from units of vermiculture/vermicompost in the form of drainage. There is no special device required to collect the vermiwash except for a tap which is fitted at the bottom of the containers where earthworms are cultured. Even during the period of normal management of the vermiculture, water is sprayed regularly to maintain adequate moisture and excess water is drained which contains some essential plant nutrients. The quality of vermiwash produced by earthworms depends on the vermicompost means source of feeding material that is used. Vermiwash, generally applied as a foliar spray act as replacement and supplement for agrochemicals and for their unique capacity to provide nutrients effectively and quickly.

7.1 VERMIWASH AND ITS COMPOSITION

Vermiwash is a honey brown colored liquid extract of organic composts, generally the wash of earthworms Present in the medium collected after the passage of water through the different layers of worm culture unit from the increased moisture content due to heat generated during vermicompost. It is a worm coelomic fluid extract containing several enzymes, plant growth hormones (IAA, Cytokinin, G A3), Vitamins, macroand micro nutrients along with excretory substances and mucus secretion of earthworms, humic acid from soil and organic waste materials which can be easily absorbed by plant tissues. It contains total solids (2448 mg/L), volatile solids (738 mg/L), silica (8 mg/L), auxin (0.98 µg/L) and cytokinin (0.68 µg/L). Dead earthworm's tissue releases nitrogen in form of nitrates-25%, ammonia 45%, organic soluble compound 3% and uncalculated material 27% which improves the nutrient quality of vermiwash. Nitrogen in vermiwash is present in the form of muscus, enzymes, nitrogenous excretory substances of worms and plant growth hormones. Vermiwash is rich in various enzymes cocktail of protease, amylase, urease and phosphatise and also microbial study of vermiwash found the presence of nitrogen fixing bacteria like Azotobacter sp., Agrobacterium sp., and Rhizobium sp., and some phosphate solubilisingbacteria.Vermiwash formed from municipal solid wastes is composed of organic matter, plant nutrient and soluble salt which increases soil nutrient and moisture content.

Table 1: Nutrient analysis of vermiwash

Parameters	Content
pH	7.39-7.5
EC	0.008±0.001
Organic carbon	0.25 ± 0.03%
Nitrogen	0.01-0.001%
Phosphorus	1.70%
Potassium	26 ppm
Sodium	8 ppm
Calcium	3ppm

Parameters	Content
Copper	0.01 ppm
Iron	0.06 ppm
Mangnesium	160 ppm
Manganese	0.60 ppm
Zinc	0.02
Total heterotrophs (cfu/ml)	1.79×10^3
Nitrosomonas (cfu/ml)	1.01×10^3
Nitrobacter (cfu/ml)	1.12×10^3
Total fungi (cfu/ml)	1.46×10^3

7.2 VERMIWASH PREPARATION

Vermiwash may be collected from the vermicompost units as a byproduct liquid extract. Whereas other method of Vermiwash preparation is followed by taking One kg adult earthworms (approximately 1000 worms) of the same species (E. euginiae) were collected and without any mixing of the casts, they were released into a 500 ml of lukewarm distilled water (37 ⁰C - 40 ⁰C) and agitated for two minutes. Earthworms were taken out and again washed in another 500 ml at room temperature (+30 ⁰C) and released back into the tanks. The agitation in lukewarm water made the earthworms to release sufficient

The effect of vermiwash was observed on the plants and soil, it was found that vermiwash seems to possess an inherent property which acts not only as a liquid organic biofertilizer which promot growth of plants and yield but also as a mild biopesticide. So, it can be used as a potent input in organic farming and sustainable crop production for both soil health and insect, pest and disease management.

quantities of mucus and body fluids. This is known as true vermiwash. Transferring into ordinary water was to wash the mucus sticking still on to their body surface and this also helped the earthworms to revive from the shock. Another method for the preparation of vermiwash, a plastic container of 15 to 20 liters capacity is required and the base of the container is fitted with tap to collect the watery worm extract. The container is filled with different succesive layers. First base layer, medium sized bricks or stones up to a height of 10-15 cm filled. Above the base layer a layer of coarse sand (up to 6 inches) and fine sand (5 inches) are spread. Introduction of locally available earthworms (Eisenia foetida) mixing with fertile soil applied. After that, a layer of partially decomposed cow dung (20-25 cm) and organic residues of 40-45 cm were poured. All the layers in the container are moistened by sprinkling water over it. Container is sprinkled with approx 2 L water per day. After 16 to 20 days preparation of vermiwash in the unit begins. Everyday about 1-2 L of vermiwash will be collected.

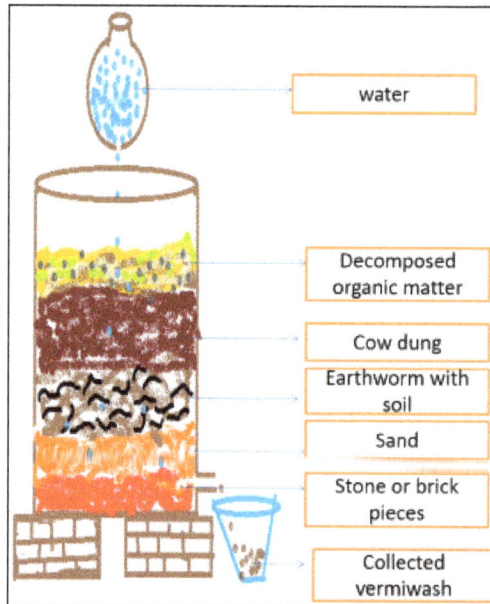

Fig 1: Preperation of vermiwash

For nutrient supplement, vermiwash (1:10) i.e @ 1 litre is generally mixed with 7-10 litres of water and the solution is sprayed on leaves of the growing crop at the evening. However, to control various pest and diseases, 1 litre of vermiwash is thoroughly mixed with 1 litre of cow urine and 10 liters of water and is kept overnight before spraying.

7.3 USES OF VERMIWASH

Vermiwash is a liquid organic fertilizer prepared from biodegradable organic wastes used both as replacement and supplement of solids and for their novel capacity to provide nutrient effectively and quickly. Being an excellent nutrient supplement, it enhances soil physico-chemical properties and maintains soil fertility in long run. It improves soil aeration, soil tilth and texture thereby reducing the compaction. It also enhances water holding capacity of soil and improves soil nutrient status of both macro and micro nutrients. It can also be added in the compost pit to hasten the degradation process. Vermiwash is known to play a major role in the plant growth and development; contribute to root initiation, root growth, plant development, promotion growth rate and improvement in crop production and improved nutrient uptake by crop and inhance nutrient content which are readily available for the plants, resulting in good crop yield. Besides serving as a plant growth promoter vermiwash has tremendous biopesticide properties. It is highly toxic against insect-pest survival and it increases disease resistance power of the crop. It is reported that mycelial growth of pathogenic fungi was inhibited at 20-30% dilution. It acts as biopesticide when it is sprayed along with 10% cow urine or neem/garlic extracts. Now a days it is also being Popularised as liquid manure.

7.4 EFFECT OF VERMIWASH ON PLANT GROWTH AND YIELD

Vermiwash is a wonderful tonic from 'farmer's friends' for boosting plant growth and yield safely, ecofriendly and economically. Vermiwash spray enhanced growth (plant height and number of leaves) and yield (number of flowers and fruits per plant) parameters when applied to brinjal plant and the vermiwash produced from animal wastes, agro/ kitchen wastes increased plant growth, flowering and productivity. Application of 100% RDF + vermiwash @100 l ha^{-1} produced distinctly higher plant height, number of primary branches, number of secondary branches, leaf area index and dry matter accumulation. It is reported that Low concentration of vermiwash is effective in bringing about seed germination and seedling growth. This could be attributed to presence of nutrients and growth promoting substances in vermiwash which showed its potentiality in seed germination and seedling vigour. Also noticed increased growth and yield response of Anthurium with the foliar application vermiwash. The seedlings of Vigna mungo, Vigna radiate and Sesamum indicum resulted in increase of growth parameters such as the root length, shoot length, number of twigs and leaves and total biomass of the plant after spraying the vermiwash of Perionyx excavates.The application of vermiwash enhanced plant height and number of leaves (56.29 cm and 6.14 days at 45 days after bud emergence), spike length and rachis (90.68 cm and 47.07 cm), number of florets (15.08), vase-life (10.02 day) number of corms m-2 (28.66), weight of corms (50.68 g) and number of cormels plant-^{1}. The treatment was also effective in reducing number of days taken to spike emergence (81.73 day). Application of vermiwash along with vermicompost also resulted in enhanced growth of plant parameters such as number of leaves, leaf length, plant height and root length to higher level.The incorporation of earthworm increased plant growth, leaf growth and root length. Vermiwash exhibited growth promoting effects on the ecomorphological characters such as plant height, length and diameter of the internode, number of leaves, leaf surface area, root length, wet and dry weight of the shoot and root of Abelmoschus esculentus. Vermicompost and vermiwash are also enriched in certain metabolites and vitamins that belong to the B group or provitamin D which also help to enhance plant growth. Combination of vermicopost and vermiwash enhanced the percentage of fats and proteins in okra. The biochemical qualities of the fruits grown in vermiwash and vermincompost indicated higher nutrient quality, which may be attributed to the presence of plant growth promoters like gibberellins, cytokinins and auxins .

7.5 Effect of vermiwash on insect pest and diseases

Vermiwash proves to have excellent bio pesticidal activities since, the plants treated with vermiwash were disease resistant and no any worms like leaf eaters were seen on the leaves and other parts of plants . Vermiwash serves as pesticide, disease curative and crop tonic and increase the yield of lab lab beans . Plants treated with vermiwash are green having vigorous growth and much more resistant to pests and disease and also .Vermiwash when applied with bio-pesticide is a preferable option for the growth, productivity as well as management of Lucinodesorbanalis infestation on brinjal crop. Combination of vermiwash and biopesticide is a superior alternative of the chemical fertilizer and pesticides for the management of Leptocorisavaricornis as well as productivity of rice crop. Since vermiwash is mild biopesticides, the combination showed synergistic effect in controlling the Leptocorisavaricornis population which ultimately enhances the productivity. The increasing concentration of vermiwash suppressed insect-pest population of tomato. Pesticidal properties of vermiwash produced using cowdung and vegetable wastes by inoculating earthworm species (E. foetida) in the laboratory to determine the effect of using the vermiwash on cowpea for mildew disease and the study revealed that usage of 20-30 percent vermiwash will cause suppression of mycelia growth of fungi. Vermicompost and vermiwash were proven to be able to control disease powdery mildew at 75.14 % rate when applied. On the other hand, combination of vermicompost, vermiwash and 10 % cow urine which said to be able to control disease at 73.37 % rate. It is also reported that vermiwash obtained from animal dung with gram bran and neem oil was also highly effective in controlling pod borer (Helicoverpaarmigera). Spray of vermiwash along with biopesticide also increased the productivity of gram crop up to 3 times with respect to control.

7.6 EFFECT ON SOIL PROPERTIES

Use of Organic formulations in agriculture could be a dynamic source to move forward soil fertility. Combination of vermicompost and vermiwash [VW+VC] recorded a significant influence on the biochemical characteristics of the soil with marked improvement in soil micronutrients and better qualitative improvement in the physical and chemical properties of the soil.The soil treated with vermicompost and vermiwash mixture had significantly inhanced soil physico-chemical properties when compared to unamended soil. Soil zinc, manganese and iron content was unaffected with the increasing application of vermiwash but Increasing the application time of the bio-fertilizers resulted in decreased zinc content to less than 1.0 mg/L and manganese content by less than 70 mg/L and increased iron content by more than 180 mg/L. However increasing the vermiwash quantity applied resulted in decreased copper content by less than 5.25 mg/L increasing the application time of the bio-fertilizers resulted in increased soil copper content by more than 8.0 mg/L. Vermiwash a liquid biofertilizer rich in the primary nutrients i.e. nitrogen (N), phosphorous (P) and potassium (K) (Nath et al., 2009 and Palanichamy et al., 2011) [11, 35] . Application of vermiwash has been reported to revitalize the soil quality (Gopal et al., 2010) [21] . It rejuvenates the depleted soil fertility and enriches available pool of nutrients, conserves moisture and natural and biological recourses. Studies revealed that application of coconut leaf vermiwash increased the crop production capacities of soil by (i) enhancing the organic carbon contents in the soil and (ii) increasing the populations of the soil microorganisms, particularly plant beneficial ones, and their activities which would have facilitated increased uptake of the nutrients by the plants resulting in higher growth and yield.

8

VERMICOMPOSTING OF AGRICULTURE WASTE

8.1 INTRODUCTION

Reprocess agricultural crop residue to vermicomposts and vermiwash provide a good option for small and marginal farmers to produce solid and liquid organic manure locally for use in their farms. Vermicomposting is a mesophilic process and should be maintained up to 32°C with the moisture content of 60-80%. Earthworms break down organic matter and leave behind castings that are an exceptionally valuable fertilizer. Vermicompost and vermiwash are well known to get better soil health and fertility as they add major and micro nutrients, organic matter, and plant growth promoting substances besides improving the soil construction. However, not much known but uniformly critical is the ability of vermicomposts and vermiwash to add considerable load of plant-beneficial microbial communities to the soil that play a chief role in augment the soil microbial diversity, nutrient mineralization, pathogen/nematode suppression and organic matter degradation resulting in better soil battle and pliability. More importantly, addition of the vermicomposts and vermiwash provides much needed food for the microorganisms in soil thereby stimulating their activities that is necessary for a soil to be fit for supporting good crop growth.While the increase in the number of Agriculture based sources supports economic growth positively, it affects the environment negatively by generating large amounts of Agriculture wastes. It is thrown away or burnt as a means of solid waste disposal. Hence there is need of development of a new method of waste disposal to benefit the agricultural as well as other industries. Agriculture waste has become a very useful substance in light of today's environmental and economic concerns. Administration of Agricultural waste has become one of the biggest problems that we are facing today. Vermicomposting is the better option to tackle with this problem. Vermicomposting is a non-thermophilic, boioxidative process that involves earthworms and associated microbes. Vermicomposting is the process of conversion of organic wastes by earthworms to valuable humus like material which is used as a natural soil conditioner. This study has attempted to assess the possibility of vermicomposting of Agriculture wastes and, in doing so, evaluated the potential of using the (Eudrilus Eugenia) to decompose Agriculture wastes.

8.2 COLLECTIONS OF COW DUNG

Animal wastes are important resources that are used to supplement organic matters and improve soil conditions. Average-size cattle produce 4 to 6 tons of fresh dung per year. However, burning of dung cakes causes serious health and environmental problems. Alternative important use of cattle dung is its conversion into compost to be used as manure in agricultural fields. For this purpose, cattle dung is heaped in the open and allowed to degrade naturally without any amendments. The cow dung has the amazing ability to attract life such as earthworms and the soil bacteria.

8.3 COLLECTION OF AGRICULTURE WASTES AND EXPERIMENTAL SETUP

The used Agriculture was obtained from different sources. Agriculture Waste and Eudrilus Eugenia earthworm species were used for the waste degradation process. Collected Agriculture wastes were chopped into small pieces. The chopped waste was mixed with cattle dung in 50: 50 ratios.

8.4 VERMICOMPOSTING TECHNOLOGY

Agriculture Waste obtained during harvesting was collected. These organic residues were shade dried for few days and cut into small pieces. It is preferable to select a composting site under shade, in an elevated level, to prevent water stagnation in pits during rains. Make small holes on the side of pits The organic residues were spread in the pit (6.0 x 1.0 x 0.6 m) up to 6^2 heights and 5 % dung slurry was uniformly distributed on the top of the organic residue sufficient to wet the surface. Over this layer another layer of organic residues was spread followed by spraying of dung slurry uniformly. This process was repeated till the spread of the organic residues 6^2 above the top of the pit. After partial decomposition of organic residues (attained in 15 days) the earthworms (Eudrilus Eugenia) were released @ 1 kg (around 1000 worms) per 1 ton organic residues in to the bed by making holes at the top of the bed on four corners and centre of the pit. Throughout the composting process, sufficient moisture was maintained i.e. at 50 percent of maximum water holding capacity of a material. Sprinkling of water should be stopped when 90 % bio-wastes are decomposed. Maturity could be judged visually by observing the formation of granular structure of the compost at the surface of the pit. Normally after 60 days, organic refuse changes into a soft, spongy, sweet smelling; dark brown compost will be ready for collection. Harvest the vermicomposts by scrapping layer wise from the top of the pit and heap under shed. This will help in separation of earthworms from the compost. Sieving may also be done to separate the earthworms, cocoons and eggs. The temperature and moisture content were maintained by sprinkling adequate quantity of water at frequent intervals. Recomposing is done in the same pit or bed. Similar to the above described. For draining of vermiwash a hole was provided which was connected to a tank with PVC pipes in order to use the vermiwash can be utilized as liquid manure. After twelfth weeks the samples were taken and were analyzed. This was studied in terms of various parameters such as pH, EC, moisture content, TOC, TKN and C: N ratio.

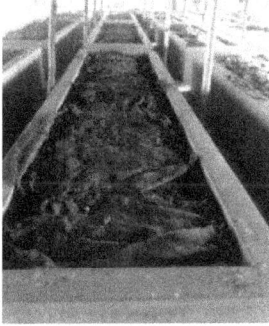

Fig. 1: Vermicompost Unit

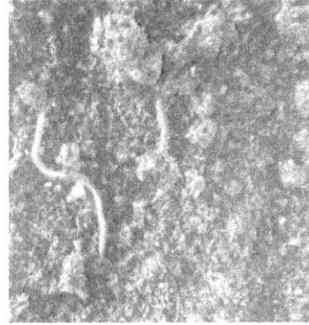

Fig.2: Movement of Earthworm in vermicompost pit

Fig. 3: Organic Manure

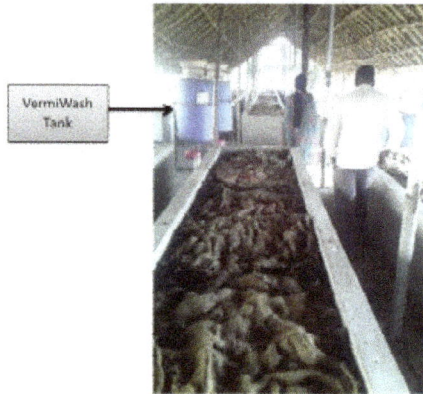

Fig. 4: Vermiwash

8.4.1 Earthworms

Multiplication of earthworms: Earthworms are bisexual, but cross-fertilization is the mode of reproduction. Adult worms, 15-21 days after copulation, lay cocoons, which look like coriander seeds. The eggs present inside the cocoon hatch into neonates in about 15-21 days. Neonates take 35-60 days to attain adulthood, which is characterized by a swollen band near the anterior part of the body. Eudrilus Eugenia, one of the species used for vermicomposting, completes its lifecycle in about 65-80 days. It leaves 400 plus cocoons in about 60 days. Vermicomposting can be tested from a small collection of pellets on the top of the beds around 45-60 days after start. This is indicative of good multiplication of worms in the beds. In about 170 days, the material is degraded completely and vermicomposts is ready for harvesting. The rate of degradation depends on the loading of worms. More number of worms, faster the degradation.

8.4.2 Vermicompost

Vermicompost is usually a finely divided peat-like material with excellent structure, porosity, aeration, drainage and moisture holding capacity. It usually contains higher levels of most of the mineral elements, which are in available forms than the parent material. Vermicompost improves the physical chemical and biological properties of soil. There is good evidence that vermicomposts promotes growth of plants and it has been found to have a favorable influence on all yield parameters of crops like wheat, paddy and sugarcane.

8.4.3 Vermiwash

Vermiwash is a liquid bio-fertilizer collected after the passage of water through a column of worm activation. It is a collection of excretory and secretory products of earthworms, along with major micronutrients of the soil and soil organic molecules that are useful for plants Vermiwash seems to possess an inherent property of acting not only as a fertilizer but also as a mild biocide an ecosystem. It contains nitrogen as nitrogenous excretory product and growth promoting hormones and essential enzymes and infuses resistance in plants. It is applied as foliar spray. This is transported to the leaf, shoots and other parts of the plants in the natural ecosystem. It contains various enzymes cocktail of protease, amylase, unease and phosphatase. These are beneficial for growth and development of plant and stimulate the yield and productivity of crops and also microbial study of vermiwash found that nitrogen fixing bacteria like Azotobacter, Agrobacterium and Rhizobium and some phosphate solublizing bacteria are also found in vermiwash.

8.5. VERMICOMPOST

A Highly Nutritive Bio-fertilizer Superior to Chemical Fertilizers Vermicompost is a nutritive plant food rich in NKP, macro & micronutrients, beneficial soil microbes like 'nitrogenfixing bacteria' and 'mycorrhizal fungi' and are excellent growth promoters. Applications of vermicomposts have been proved effective to enhance growth and yield of various plants. Vermicompost also contain enzymes like amylase, lipase, cellulase and chitinase more significantly, vermicomposts contains 'humus' which makes it markedly different from other organic fertilizers. It takes very long time for soil or any organic matter to decompose to form humus while earthworms secrete humus in its excreta. Without humus plants cannot grow and survive. The humic acids in humusare essential to plants in four basic ways – 1). Enables plant to extract nutrients from soil; 2). Help dissolve unresolved minerals to make organic matter ready for plants to use; 3). Stimulates root growth; and, 4). Helps plants overcome stress. Vermicompost can be used for all crops: agricultural, horticultural, ornamental and vegetables at any stage of the crop.

Fig. 5: Growth of panicle in SRI rice with organic manure

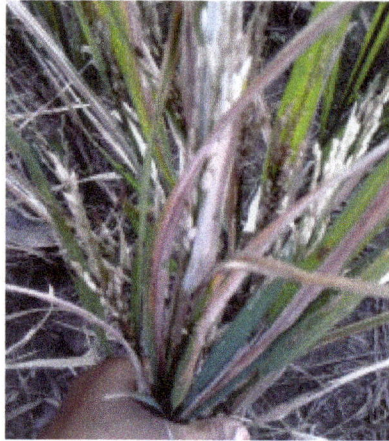

Fig. 6: Growth of panicle in rice with chemical fertilizer

8.6 EFFECTS OF VERMICOMPOST

Earthworms play an important role in maintaining soil fertility through vermicomposting. In the present study, The lowering of pH due to production of CO_2 which was an acidic gas and when it came in contact with water it might had formed carbonic acid, due to which pH had decreased. Generally there was an important decrease in EC, which is superlative for plant growth. With a low EC, the Organic fertilizers releases the mineral salts gradually, which is adequate for plant escalation. There was a noticeable reduction in the TOC and TOM in the final vermicompost prepared from waste using EudrilusEugenia .It is due to the microbial respiration.

Table 1: Physicochemical analysis of Agriculture waste based Vermicompost

S. No	Parameters	Days		
		0	35	85
1	PH	8.2	7.8	7.4
2	Electrical conductivity (EC)	3.50	3.25	3.01
3	Total Kjeldahi Nitrogen (TKN)	0.46	0.72	0.11
4	Total Phosphorus (TP)	0.04	0.12	1.11
5	Total Potassium (TK)	0.38	0.46	1.20
6	Total organic carbon (TOC)	21.3	12.8	8.6
7	Carbon Nitrogen Ratio (C:N)	50.6	26.4	11.8

The N content percentage increase might instigate from the addition of nitrogen through the earthworm itself in the form of mucus, nitrogenous execratory substance, growth stimulating hormones and enzymes. Phosphorus increased by the closing stages of the process owed to the mineralization of organic matter. Increase in K possibly due to the direct action of earthworm guts and indirectly by the simulation of micro flora. Moreover, the Increase in earthworm population might also be attributed to the C: N ratio decreasing with time. Decline of C: N ratio to less than 20 indicates an advanced degree of organic matter stabilization and reflects a satisfactory degree of maturity of organic waste.

Table 2: Impact of vermicomposting on weight loss of organic substrate in the presence of Earthworms

Type of waste	Earthworm	Initial weight of substrate of mixed three materials (gm.)	After 85 days, final weight of vermicompost (gm.)	Loss of organic waste during vermicompostin (%)
Cow dung + Agriculture waste	Eudrilus Eugenia	2500	900	83

Earthworms rapaciously feed on organic wastes and while employing only a small portion for their body synthesis they excrete a large part of these consumed waste materials in a half-digested form. Since the intestines of earthworms' harbor wide ranges of microorganisms, enzymes, hormones, etc., these half-digested materials decompose rapidly and are transformed into a form of Vermicompost within a short time.

Table 3: Earthworm Biomass during Vermicomposting Period

Waste material	Earthworm	Earthworm Number in Days		Body Weight of Earthworm	
		0	90	0	90
Agriculture waste	Eudrilus eugenia	1000	1960	10.60	38.31

8.7 APPLICATION OF VERMICOMPOST &VERMIWASH TO CROPS

Vermicompost can be used for all crops such as agricultural, horticultural, ornamental, and vegetable etc.

Table 4: General rate of Vermicompost application in different crops

Crops	Rate
Field Crops	3-5 t/ha
Vegetable crops	5-7 t/ha
Fruit crops	3-5 kg/tree
Flower crops	100 g/pot
Nursery bed and lawns	1-2 kg/m2

Vermicompost improves soil aeration because they do not pack together when mixed in soil. This in turn promotes rapid plant growth. Earthworm castings improve the soil's drainage, reducing waterlogged soil and root rot. The soil's water retention capacity also improves because Vermicompost contains absorbent organic matter that hold only the necessary amounts of water needed by the roots. But generally, Vermicompost is recommended for high value vegetables and fruit crops.

Agricultural waste is no hesitation organic waste and is purely biodegradable. Hence it can be used for composting. Cow dung is easily accessible material in rural area. If it is added with the waste microbial activity increases and oxidation of organic matter takes place in faster rate and stabilized within shorter period. Also cow dung adds nutrient to the compost. Thus increases the quality of final compost. This study concludes that the vermicompost of Agriculture wastes can be utilized as an organic fertilizer instead of being disposed in landfills. The results also showed that the smaller the size of the shredded waste, the more rapid the decomposition process. The success of E. Eugenia to Vermicompost Agriculture wastes was also studied, and it was found that, recycling is truly an ecofriendly technology through which one can convert all organic waste into a product which is rich in nutrient content and can replace chemical fertilizer. Today we are spending a lot of money to buy chemical fertilizers and pesticides. Most of these are petrochemical based, which are not available indefinitely and are becoming more and more expensive with more foreign exchange required. In addition to high price of chemical fertilizers, it has negative impact on the environment. Environment gets highly polluted due to excessive use of chemical fertilizers and pesticides. The small farmers cannot afford these chemical fertilizers because the soil needs more and more of these chemicals. On the face of this, vermicomposts technology will save our environment. In turn, we will have much better sustainable agriculture. Hence, farmers welcome this type of enterprise.

9

ORGANIC WASTE MANAGEMENT BY EARTHWORM

9.1 INTRODUCTION

The habitations of human beings and animals generate huge load of organic wastes and their decomposition products affect the quality of soil, air and water. Much of the biosolid wastes are highly infectious as they contain an array of pathogenic microorganisms. Their disposal into the environment without prior disinfection causes health and environmental risks. The animal and municipal waste change the soil character, including pH, bulk

Vermicomposting is a biotechnological process involved by earthworm; the natural bioreactors playing an essential role in the breakdown of organic matter and maintaining soil fertility. The worms involved recycling of organic waste and enhanced plant growth. The importance of vermicompost is further enhanced as it has simultaneously other benefits; excess worms can be used in medicines and as protein rich animal feed. Finally, we conclude that vermicompost reduced the pesticide application, low pest infestation, reduction of irrigation frequency and pesticide free high yield.

density, conductivity, water holding capacity and increased the organic carbon content. The management of biosolid wastes is gaining importance for not only providing clean and healthy environment but also enhancing the primary productivity through soil quality improvement. Composting is the widely accepted process for the recycle in organic wastes; the direct land application of raw wastes or poorly stabilized materials caused toxicity and pathogenicity towards land. Composting is the biological transformation of organic matter into a well-stabilized product through the fast succession of microbial populations under aerobic conditions. The process results in mineralization of organic matter into carbon dioxide and transformation into humic substances.

Vermicomposting is a biotechnological process, in which organic materials converted as valuable product by earthworms. The nutrient profile of vermicompost is higher than traditional compost. The vermicompost alters the soil fertility in different ways, such as better aeration, porosity, bulk density, water holding capacity, pH, electrical conductivity, nitrogen, phosphorous and potassium content. The application of the vermicompost is enriches the soil microorganism, plant growth (size of leaf, height, width and weight) and nutrient content of the yield. The high concentrations of vermicompost may delay plant growth due to the concentration of soluble salts. As a result, vermicomposts should be applied at required quantity to produced higher yield. In this overview describe about the organic waste management, vermicompost, earthworm's species and economical importance of the vermicompost.

Most of the Indian cities are threading by environmental problem due to solid waste. The solid waste creates many problems in raining season by blocking the running water. Many ways solid waste was managed viz., incineration, composting, gasification, refuse derived fuel (RDF). The organic solid was pulverization and converted into vermicompost, used as fertilizer with economical worth of 9.36 lakhs/year. The household waste of food, paper, vegetable and garden (grass and leaves) with cow dung were converted as vermicompost by Eudrilus sp.; the compost having rich benefi cial microbial community of bacteria, fungi, actinomycetes, Pseudomonads, P- Solubilizers and N_2 Fixers. Indiscriminate uses of synthetic chemicals lead to many problems in agroecosytem and affect the non-target organism. Agricultural industry waste such as sheep manure, pomegranate peels, mushroom, chopped corn, sugar beet pulp and sawdust were used as raw material for vermibed; and the obtained compost reduced the electrical conductivity, raising pH, and NPK of treated land and act as fertilizer.

Vermicomposting, is a bio-oxidative process in which earthworms interact intensively with micro-organisms in the decomposer community, accelerating the organic matter by stabilization with modifi ed physical and biochemical properties. Vermicomposting differs from conventional composting because the organic material is processed by the digestive systems of earthworms. The digested casts can be used to improve the fertility and physical characteristics of soil. In this process, the earthworms actively participate in the degradation of organic matter by physical and biochemical action. Physical participation in degrading of organic substrates results in fragmentation, for increasing the surface area to action and aeration. Conversely, biochemical changes in the degradation of organic matter are carried out by microorganisms through enzymatic digestion, enrichment by nitrogen excrement and transport of inorganic and organic materials. The earthworms contribute significantly in the recycling of organic waste and production of organic manure with high humic contents, which are helpful to maintenance the soil structure, aeration and fertility. The bioactive substances present in the humic acid fertilizer can enhance physiological metabolism, growth, yield, seed germination etc., while these features are absent in ordinary fertilizers. Applying humic acid fertilizer can also effectively increase the anti-drought and ant frigidity potential of crops, and prevent underground plant diseases, insect pests and pathogenic bacteria. The present reviews describe about various aspects involved in vermicomposting of organic waste by different species of worm.

9.2 BIOLOGY OF EARTHWORM

Earthworms are invertebrates of agro-ecosystem, belonging to the family lumbricidae, both male and female reproductive organs present in single earthworm, hermaphrodites. At the time of eggs laying, the sexually mature worms contain a distinctive epidermal ring-shaped area called, clitellum, which has gland cells that secrete material to form a viscid, girdle like structure known as cocoon. The number of fertilized ova in each cocoon have 1-20 lumbricid worms. There were about 3000 species of earthworms distributed all over world and about 384 species were reported in India.

9.3 CHARACTERISTICS AND PROCESS OF VERMICOMPOSTING

 It is an aerobic, bioxidation and stabilization non-thermophilic process of organic waste decomposition that depends upon fragments by earthworms, mix and promotes microbial activity. It is a peat like material with high porosity, aeration, drainage, water holding capacity and microbial activity. It also enhanc the resistance of plants against pests and diseases. Earthworms can serve as "nature's plowman" and form nature's gift to produce good humus, which is the most precious material to fulfil the nutritional needs of crops. Organic waste pollution was increased by day-to-day activities and other side shortage of organic manure, in this connection earthworms were used for conversion of organic waste into vermicompost. Vermicomposting is a powerful tool for bulk reduction of waste as well as pathogen free vermicompost. The vermicompost reduced the cadmium content of the soil and enrich the soil by maintain the pH, P, K, Na, Ca and microorganism. Some of the waste was converted as methane by anaerobic; it leaves global warming, in this case, earthworm was used for organic waste management without environmental pollution. Tea waste was mixed with soil in different ratio, for conversion as compost by using E. eugeniae. The compost provides rich nutrient with microbial content called as fertilizer. The accumulation of organic waste, threat to the environment in all the continent, the waste would be managing the sustainable way by using anaerobic conditions without affecting ecosystem. It may use for biogas production and energy management for day-by-day usage.

Vermicomposting is an important technique of converting organic waste into nutrient rich compost by earthworms without compromising the population of beneficial bacteria. In recent years, many researchers concentrated to study about vermicompost for sustainable agriculture, the organic waste was converted as vermicompost, the compost have high content of NPK, Carbon, nitrogen, beneficial microorganism and growth hormones. Vermicompost is process of non-thermophilic, bio-oxidative process with help of earthworm and microbes. Compost enhanced the soil fertility by soil biodiversity richness, water holding capacity and growth regulation hormones.

Compost has high microbial content, required pH, organic matter, moisture content, nitrogen, phosphorous, potassium and Carbon: Nitrogen ratio (C: N) value for agricultural practice. Solid waste management is a big issue in our country; the waste was converted into useful agricultural fertilizer by using earthworm.

The livestock excretes was managed by sustainable way called as vermicompost. Without earthworm, the composting will take energy losses (used number of man power), while using earthworm its breakdown the manure in very fast and would be recyclable; the vermicompost having high amount of nutrient and microbial biomass. Compost is a cost important tool for eco-friendly waste management. The obtained vermicompost having increased NPK, calcium, magnesium, sodium and required EC and C/N ratio. Eisenia fetida was used for biodegradable solid waste management and it is ecofriendly waste management without producing any heat. As suggested that small scale (individual house) waste management is easiest way to solid waste management and economically viable. Also, the resulted of raw waste was lower than compost pH. The carbon, phosphorus, carbon:nitrogen ratio of the compost was higher than raw waste.

Solid waste management is one of the big environmental issues. The residual and animals waste were polluting the aquatic and terrestrial ecosystem. A quantity of biodegradable solid waste categorized. Such waste modifi ed into environmental safely products by microbial composting, vermicomposting, biogas plant, etc; Vermicompost induced, germination, flowering and fruiting the plants earlier than the control (without vermicompost). The increased micro flora (bacterial and fungal colonies) and NPK content was increased in treatment, when compared to the control. Vermicompost is a simple, cost effective, low maintenance, easy method of waste management. In India reusable organic biomass of 0.7 and 0.8 million tonnes/year. The author stated that potential use areca nut and cocoa waste were managed by vermi technology and without environmental effects; it having higher nutrient than compost; also, it induced the soil fertility.

9.4 WASTE STABILIZATION BY VERMICOMPOSTING

A greenhouse trial was conducted with vermicompost from 1. Raw dairy manure with tobacco residue, 2.yard leaf, 3. Sewage sludge + rice hull, 4. Sewage sludge + yard leaf, and raw dairy manure were evaluated in tomato seedling; all the treatment produced a significant growth than control. The kitchen waste was thermo composted for 9 day and converted as vermicompost within 2.5 months. The house hold and market vegetable waste were mixed and finally add 0, 10, 20 and 30% of sewage sludge. The mixture was aerobically decomposed for 20 days and E. eugeniae was released. Within 30 days, end product was obtained when compared to windrow composting (80 days).

The agriculture waste and cattle manure with different ratio was prepared as vermibed such as 1. Equal weight of (Penniseum typhoides and Sorghum vulgare) + sheep manure

(1:2 ratio), 2. Vignaradiata+ Triticum aestivum+ cow dung (1:1:2 ratio), 3. Mixed all the plant + cow dung (1:1 ratio) 4. Cattle shed manure for renewal energy by vermicompost. All the compost has rich N (97.3% to 155%), P (67.5% to 123.5%), K (38.3% to 112.9%), and Ca (23.3% to 53.2%), and decrease in organic C (20.4% to 29.0%). The temple organic wastes with cow dung and biogas digester slurry were decomposed for a period of 30 day at 30° C. After, digestion E. eugeniae was introduced at 25° with pH 8.0 and moisture content (80%) should be optimum for high nutrient yield of vermicompost.

E. fetida and L. mauritii were used for poultry waste management. Both species produced compost 3rd day onwards. E. fetida was performed well when compared to L. mauritii; but there were no significant changes in the soil and soybean husk and papaya waste were studied in different ratio for waste management with E. eugeniae. Among the different ratio 1:1 ratio was best combination for vermicompost. Zoo waste where mainly animal dungs, garbage and litter; converted as compost by using earthworm, the composting period was minimum than traditional compost. African earthworm E. eugeniae used as solid waste management. Paper waste contain rich carbon but need nitrogen for that cow dung was used as mixture in different ratio a. 1:1 (paper+ cow dung), b- 2:1 (paper + cow dung), c- 3:1 (paper + cow dung), among the combination 1:1 ratio was best one, the compost was collected up to 65 days excluding 15 days pre digestion. The olive wastes were converted as vermicompost by using E. andrei. The obtained vermicompost having increased NPK, calcium, magnesium, sodium and required EC and C/N ratio.

Solid waste was managed by earthworm. The waste called as gold, when it was converted as vermicompost and will not produce any kind of risk to the environment. Different ratio of verminbed were T1, soil + cow dung, (0.5:1), T2, soil+vegetable+fruit waste (1:1), T3, soil+agricultural waste+cow dung (1:1) and T4, soil+paper waste+cow dung (1:1). The vermicompost showed increased potash and nitrogen content. It is stated that 5-75% of organic solid waste collected from religious area of Patna, Bihar, India; it converted as valuable materials by using E. fedita and E. eugeniae. The result showed that NPK were increased in duration was increased also C/N ratio, carbon% were decreased increasing duration of composting period.

Kitchen waste, with combination of cowdung, were predigestion for a period of 15 days. The best compost was obtained in kitchen waste + cowdung (1:1), then Kitchen waste+cowdung (2:1) and Kitchen waste+cowdung (3:1) ratio. The pH value was reduced when duration of the compost collection day increased (0- 60 day), total nitrogen (%), phosphorus (%), potassium (%) and carbon: nitrogen ratio was increased considerably. It is observed that organic solid waste management is major problem in developing countries. Poor management of solid waste is called as environmental pollution. In India most of pilgrims, generated large quantity of waste during the festival. The flower waste was high in some major temple of Jaipur, Rajastan, India. The fl ower waste was converted as vermicompost with mixture of cattle dung by using E. foetida. The waste was converted as vermicompost with 50% of weight was reduced during the vermicompost process. The compost was at 25°C, 8.0 pH, 1-2 mm particle size, moisture 60%, bulk density was acceptable limit. Also compost reduced EC, C: N ratio, C: P ratio and increase in N, P.K, Ca, Mg, and sulphur. The vermicompost used as fertilizer for tomato plant cultivation it enhanced the growth (stem diameter, height, leaf number, length of roots, yield/plant).

9.5 POTENTIAL APPLICATIONS OF VERMICOMPOST IN PLANT GROWTH

Vermicompost (cattle manure) was studied in their efficacy on Petroselinum crispum, the result indicated that vermicompost enhance the size of leaves, plant height and yield. Earthworm play, a vital role in organic waste management by vermicomposting; the vermibed was prepared in different combination by using sewage sludge amended with hazelnut husk and cow dung. After preparation, the E. fetida was introduced in to vermibed and their compost was studied on Triticum aestium; found that all treatment induced the growth and yield of tested plants when compared to control.

Every population released considerable amount of organic waste, it is dumping in landfills (burn or river systems). The river was polluted by plant and large amount of market waste daily. While, the waste was converted as vermicompost (30% yield), huge volume was degreased. For the reason earthworm is a main technology transfer for bio-waste into valuable materials. Also, it contains beneficial microorganisms (Actinomycetes, Azotobacter, Nitrobacter, Nitrosomonas and Aspergillus) for plant productivity. The effect of vermicompost, chemical fertilizer, 50% manure +50% fertilizer and control on rice field. They found that vermicompost showed good growth and provided maximum nutrient to tested plants.

The increasing population, there is need for fertile lands to cultivation. Mentha arvensis was cultivated under salt stressed conditions in both controlled and fi eld conditions. The fungi Glomus aggregatum and Exiguobacterium oxidotolerans with vermicompost improved plant growth. Also concluded multi-microbial inoculations together with vermicompost as efficient biofertilizers for M. arvensis cultivation. Vermicompost had complex effects on the antioxidant enzyme activities of plants; when it was grown under high salinity. The organic waste of industrial sewage sludge and municipal solid waste compost were mixed with cultivated soil. The both treatments increased the organic matter, microbes, but decreased waster holding capacity in industrial waste, in case of the yield industrial waste was higher than municipal waste treated soil.

Waste materials were digested by anaerobic condition then converted as vermicompost by using earthworm. While vermicompost applied in the fi eld, it reduced the irrigation frequency and induced the plant growth. The vermicompost from sewage sludge, wood chips mixture with biochar, induced the higher reproductive rate of earthworm (cocoon,

juveniles) and reduced the Zn and cd. As humic acid rich in vermicompost, is induced the plant height, fresh weight, dry weight. Also, maximum number and density of microbes (bacteria, fungi) were recorded when compared to chemical fertilizer. Different species of earthworms was used for production of vermicompost; and is powerful biofertilizer in sustainable agriculture with reduction of chemical agrochemicals. The worms were involved in waste manage ment by and recycling of organic waste.

10

ROLE OF VERMICOMPOSTING IN IMPROVEMENT OF SOIL NUTRIENTS AND AGRICULTURAL CROPS

10.1 INTRODUCTION: AN OVERVIEW

Composting, generally defined as the biological aerobic transformation of an organic byproduct into a different organic product that can be added to the soil without detrimental effects on crop growth. In the process of composting, organic wastes are recycled into stabilized products that can be applied to the soil as an odorless and relatively dry source of organic matter, which would respond more efficiently and safely than the fresh material to soil organic fertility requirements. The conventional and most traditional method of composting consists of an accelerated biooxidation of the organic matter as it passes through a thermophilic stage (45° to 65°C)

Earthworm's activity influences the rate of soil turnover, mineralization and humification of soil organic matter. Improvement in the consistency of soil texture with a concomitant increase in porosity, infiltration and soilwater retention are other characteristics of worm-worked soils. There are multiple benefits of vermitechnology; low cost production of biofertilizer, environmental management of solid wastes and agricultural residues, enhanced soil productivity, tastier quality food, among others. Vermitechnology also aids in the reduction of soil salinity, soil erosion with less runoff and wasteland development. From this present review, it is concluded that the organic wastes are effectively recycled by microorganisms followed by earthworms and plays a major role in the development of growth and yield of agricultural crops. The nutritive value of compost material is high and the composting process effectively converts the waste product into useful by-product.

where microorganisms liberate heat, carbon dioxide and water. However, in recent years, researchers have become progressively interested in using another related biological process for stabilizing organic wastes, which does not include a thermophilic stage, but involves the use of earthworms for breaking down and stabilizing the organic wastes.

Composting is a biotechnological process by which different microbial communities convert organic wastes into a stabilized form. During the process, thermophilic temperatures arise because of the heat released due to biological activity. These temperatures are responsible for pathogen inactivation. Composting is an aerobic process that requires oxygen, optimal moisture and enough free air space and C/N ratio within certain limits. The treatment by composting leads to the development of microbial populations, which causes numerous physicochemical changes within mixture. These changes could influence the metal distribution through release of heavy metals during organic matter mineralization or the metal solubilization by the decrease of pH, metal biosorption by the microbial biomass or metal complexation with the newly formed humic substances (HS) or other factors.

The tremendous increase in population, urbanization, industrialization and agricultural production results in accumulation quantities of solid wastes. This has created serious problem in the environment. In order to dispose this waste safely it should be converted effectively. This is achieved by bio-composting and vermicomposting of farm, urban and agro-industrial waste. It is being increasing realized that composting is an environment friendly process, convert wide variety of wastes into valuable agricultural inputs. Compost is excellent source of humus and plant nutrients, on application of which improve soil biophysical properties and organic matter status of the soil. This present review focused on vermicomposting and its importance in improvement of soil nutrition and agricultural crops. This review assesses the following topics: vermicomposting, raw materials of vermicomposting, microbiology of vermicomposting, effect of vermicompost materials in agriculture and physico-chemical properties of soil, and importance of vermicompost. Recycling organic wastes through vermicomposting is being considered as an economically viable solution. Earthworms are considered as natural bioreactors while proliferate along with other microorganisms and provide required conditions for the biodegradation of wastes.

10.2 VERMICOMPOSTING

Earthworms are often referred to as farmer's friends and natures ploughmen. Earthworms are extremely important in soil formation, principally through their activities in consuming organic matter, fragmenting and mixing it intimately with mineral particles to form aggregates. During their feeding, earthworms promote microbial activity greatly, which in turn accelerates the breakdown of organic matter and stabilization of soil aggregates. The ability of some earthworms to consume a wide range of organic residues such as sewage sludge, animal wastes, crop residues, and industrial refuse has been fully established. In the process of feeding, earthworms fragment the waste substrate, enhance microbial activity and the rates of decomposition of the material, leading to a composting or humification effect by which the unstable organic matter is oxidized and stabilized. The end product, commonly termed vermicompost and obtained as the organic wastes pass through the earthworm gut, is quite different from the parent waste material.

Vermicomposting is a simple biotechnological process of composting, in which certain species of earthworms are used to enhance the process of waste conversion and produce a better end product. Vermicomposting differs from composting in several ways. It is a mesophilic process, utilizing microorganisms and earthworms that are active at 10–32°C (not ambient temperature but temperature within the pile of moist organic material). The process is faster than composting because the material passes through the earthworm gut, a significant but not yet fully understood transformation takes place, whereby the resulting earthworm castings (worm manure) are rich in microbial activity and plant growth regulators, and fortified with pest repellence attributes as well in short, earthworms, through a type of biological alchemy are capable of transforming garbage into 'gold'.

Vermicompost are finely divided peat-like materials with high porosity, aeration, drainage, and water-holding capacity. They have a vast surface area, providing strong absorbability and retention of nutrients. Vermicompost contain nutrients in forms that are readily taken up by the plants such as nitrates, exchangeable phosphorus, and soluble potassium, calcium, and magnesium. Decomposition of various organic substrates (kitchen waste, agro-residues, institutional and industrial wastes including textile industry sludge and fibers) into valuable vermicompost has been extensively studied using an exotic earthworm species (epigeic Eisenia foetida). Tests have also been conducted combining thermo-composting and vermicomposting to improve efficiency and compost quality.

As advocate the integrated use of organic and inorganic nutrient sources with effective microorganisms (EM) for improving crop yield. The effects of earthworm processed sheep-manure (vermicompost) on the growth, productivity and chemical characteristics of soybean straw (Glycine max L. Merril.), wheat straw (Triticum aestivum L.), maize stover (Zea mays L.), chickpea straw (Cicer arietinum L.), city garbage and greenhouse tomatoes (Lycopersicum esculentum) has also been studied. Earthworm species such as Eudriluseugineae are voracious feeders of organic wastes, and their presence has been found to reduce the time required for composting.

10.2.1 Raw Materials for Vermicomposting

The residues like sugarcane trash, press mud, sugar factory effluent, broiler ash, spent wash, etc, should be bio processed and added to the soil, to complete their natural cycle. Bicycling of these residues through vermiculture biotechnology reduces the use of chemical fertilizers derived from non-renewable sources. "BIOAGRO" compost was produced from the city garbage. By the addition of neem cake, rock phosphate and gypsum in small quantities to this compost "BIOAGRORICH" compost were made. Organic wastes such as poultry manure, cattle dung, pig manure as well as agricultural waste like sugarcane trash were fed to earthworm to hasten the process of decomposition. As reported that the waste consists of decomposable organic matter with high carbon nitrogen ratio. Hence the organic matter wastes are composed by vermicomposting process in order to convert the organic waste into bio-compost. Also reported that the vegetable market waste is leftover and discarded rotten vegetables fruits and flowers in the market. This urban waste can be converted to a potential plant nutrient enriched resource compost and vermicompost that can be utilized for sustainable land restoration practices.

10.2.2 Microbiology of Vermicomposting

Due to inoculation of microorganisms the period of composting was reduced by about 4 weeks. The results also indicate that by utilizing mesophillic cellulolytic fungi, the process of composting a high C/N homogenous material can be accelerated and the quality of the resulting composting can be improved. Various studies also indicated the possibility of augmenting the quality of compost through inoculation with Azotobacter and phosphate solubilizing microorganisms in the presence of rock phosphate.

The role of earthworms as vectors of beneficial soil bacteria and their capacity of influence the population dynamics and impact of microorganisms on soil and plants was studied. Actinomycetes and bacteria (both celluloolytic and lignolytic) which are important in waste degradation increase exponentially along the entire length of the tabular bioreactor.

The gut isolates included the Actinomycetes, Streptomyces lipmanii and the oxalatedegrading bacterium Pseudomonas oxalaticus and anaerobes have not been enumerated from the worm gut but several nitrogen fixers (Clostridium butyricum, Clostridium beijerinkii and Clostridium paraputrificum) have been isolated from Eisenia foetida casts, microbial growth was limited by the amount of available carbon immobilization of phosphate in earthworm casts is probably caused by mainly abiotic processes, carbon mineralization by soil microflora fertilizer with glucose and phosphorous was limited by nitrogen, except in freshly deposited casts.

The gut of earthworm behaved as an epigenic/anecic species in sugarcane fields, where it seems to feed on decayed sugarcane liter and deposits its casts on the soil surface. The digestive enzymes and intestinal microflora of earthworms seem to play an important role in digestive of soil organic matter, the various enzymes viz., amylase, cellulose, xylanase, cellbiose, endonuclease, acid phosphatase and their activities in the gut of the two selected earthworms Eudriluseugenie, Eisenia foetids. The changes in bacterial community play a major role during vermicomposting. In addition to bacteria, fungi especially cellulolytic fungi also play an important role during vermicomposting. Population of cellulolytic fungi was found to be increased during vermicomposting of different organic wastes. Cellulase produced by these fungi plays a major role in decomposition of cellulolytic materials of organic wastes.

10.3 EFFECT OF VERMICOMPOST MATERIALS IN AGRICULTURE

Vermicomposting is a process of biotransforming and stabilizing organic materials (often waste) into humus by the combined activity of earthworms and microorganisms. Earthworms excrete partially digested materials, known as vermicasts or castings, which are more homogeneous in composition than the source material, have reduced levels of contamination, and contain elevated levels of plant growth regulators or symbiotic microbes and organic acids such as humic and fulvic acids. Vermicomposting refers to production of compost by growing/ breeding earthworms as these worms in the process of feeding on waste cause biooxidation by relentless turning, fragmentation and aeration of waste by devouring resulting in homogeneous and stabilized humus like product which is an ideal nutrient for plants thus used as manure. Vermicomposting of biodegradable Municipal Solid Waste and household waste is in vogue in many places and instances but there is no available literature on use of vermicomposting technique for treatment and disposal of infected biomedical waste. There is an emerging commercial trend of aerobically incubating an extract of compost with a carbohydrate and a protein source, producing a microbially enhanced liquid. Known by the agricultural sector as 'compost teas' in the current study this microbially enhanced product is termed "Compost Extract" or "CE" in short. Compost extract contains nutrients extracted from compost and thus contributes directly to plant nutrition, and also contains organic matter, improving soil structure and water holding capacity by building soil aggregates. Composting and vermicomposting are quite distinct processes, particularly concerning the optimum temperatures for each process and the types of microbial communities that predominate during active processing (i.e. thermophilic bacteria in composting, mesophilic bacteria and fungi in vermicomposting). The wastes processed by the two systems are also quite different. Also reported that vermicomposts have a much finer structure than composts and contain nutrients in forms that are readily available for plant uptake and the production of plant growth regulators in the vermicomposts. Therefore, they hypothesized that there should be considerable differences in the performances and effects of composts and vermicomposts on plant growth when used as soil amendments or as components of horticultural plant growth media.

Application of vermicomposting in combination with NPK fertilizers resulted in higher content of total nitrogen compared to FYM in combination with NPK fertilizers or control. It also resulted in higher content of phosphorus significantly. The casting by earthworms was seen to improve, the soil organic matter and nutrient status, by recycling available nutrients especially N, P, K, Ca and Mg. Application of coir dust coir pith into soil contributes 20.7 kg N, 10.5 kg, P2O5 and 30.8 kg K2O ha annually. Coir pith being a rich potash source also helps to retain moisture in the soil for a long time.

10.4 EFFECT OF VERMICOMPOST ON PHYSICO-CHEMICAL CHARACTERISTICS OF SOIL

The composted organic wastes exert variety of physical, chemical and biochemical influences upon the soil faking the soil a favourable substrate for plant growth. It maintains the soil in a proper homeostaitc state. It also removes excessive amounts of heavy metals such as copper and lead and there by served as a means of detoxification. As reported that there was a slight decrease of pH due to the organic acids released during the decomposition of the various farm wastes. The EC value was also altered by the organic waste application into the soil. There was a significant increase in the available N status due to the application of the various farm waste materials. The available P status was also significantly increased by the application of the various farm waste materials. The application of organic wastes into soil has considerably increased the available K status also.

Application of vermicomposting in combination with NPK fertilizers resulted in higher content of total nitrogen compared to FYM in combination with NPK fertilizers or control. It also resulted in higher content of phosphorus significantly. The casting by earthworms was seen to improve, the soil organic matter and nutrients status, by recycling available nutrients especially N, P, K, Ca and Mg. Application of coir dust coir pith into soil contributes 20.7 kg N, 10.5 kg, P2O5 and 30.8 kg K2O ha annually. Coir pith being a rich potash source also helps to retain moisture in the soil for a long time.

Nutrient composition and physico-chemical parameters are subjected to greater changes due to activity of earthworms in food substrates, while mineralization of waste substrates is accelerated by passing of ingested food through gut of earthworms, thus stabilizing NPK contents in plant available form. The nutrient contents and physico-chemical parameters of vermicompost samples obtained in the present study looked optimum and are considerable to such earthworm mediated compost.

The application of vermicompost increased the growth and yield of paddy besides increasing the levels of total nitrogen, available phosphorous and potassium and micronutrients in the soil. The fertilizing effect of earthworm casts depends on microbial metabolites, mainly growth regulators. Earthworm casts promoted the root initiation and root biomass. The chemical fertilizer application along with vermicompost increased the nutrients uptake and the net production of wheat and sugarcane. Vermicompost is superior to normal compost in increasing the growth of cardamom seedlings. Significant increase in the yield due to saw dust was observed compared to nitrogenase fertilizers application alone. The general vigor of the crop was more with profuse tillering under saw dust compost amendments while with nitrogenous fertilizer alone spans or no tillering was observed.

Vermicompost contains plant growth regulators and other plant growth influencing materials produced by microorganisms. Vermicompost a by-product of earthworm mediated organic waste recycling is rich in plant nutrients and growth promoting substances. Having shown to promote and sustain crop yields.

10.5 IMPORTANCE OF VERMICOMPOST

10.5.1 Source of Plant Nutrients

Earthworms consume various organic wastes and reduce the volume by 40–60%. Each earthworm weighs about 0.5 to 0.6 g, eats waste equivalent to its body weight and produces cast equivalent to about 50% of the waste it consumes in a day. These worm castings have been analyzed for chemical and biological properties. The moisture content of castings ranges between 32 and 66% and the pH is around 7.0. The worm castings contain higher percentage (nearly twofold) of both macro and micronutrients than the garden compost. Soil available N increased significantly with increasing levels of vermicompost and highest N uptake was obtained at 50% of the recommended fertilizer rate plus 10 t ha^{-1} vermicompost. Similarly, the uptake of N, phosphorus (P), Potassium (K) and magnesium (Mg) by rice (Oryza sativa) plant was highest when fertilizer was applied in combination with vermicompost.

10.5.2 Improvement of Plant Growth and Yield

Vermicompost plays a major role in improving growth and yield of different field crops, vegetables, and flower and fruit crops. The application of vermicompost gave higher germination (93%) of mung bean (Vigna radiata) compared to the control (84%). Further, the growth and yield of mung bean was also significantly higher with vermicompost application. Likewise, in another pot experiment, the fresh and dry matter yields of cowpea (Vigna unguiculata) were higher when soil was amended with vermicompost than with biodigested slurry.

10.5.3 Reduction in Soil C:N Ratio

Vermicomposting converts household waste into compost within 30 days, reduces the C:N ratio and retains more N than the traditional methods of preparing composts. The C:N ratio of the unprocessed olive cake, vermicomposted olive cake and manure were 42, 29 and 11, respectively. Both the unprocessed olive cake and vermicomposted olive cake immobilized soil N throughout the study duration of 91 days. Cattle manure mineralized an appreciable amount of N during the study. The prolonged immobilization of soil N by the vermicomposted olive cake was attributed to the C:N ratio of 29 and to the recalcitrant nature of its C and N composition. The results suggest that for use of vermicomposted dry olive cake as an organic soil amendment, the management of vermicomposting process should be so adjusted as to ensure more favorable N mineralization and immobilization.

10.5.4 Role in Nitrogen Cycle

Earthworms play an important role in the recycling of N in different agro-ecosystems, especially under jhum (shifting cultivation) where the use of agrochemicals is minimal.

As reported that during the fallow period intervening between two crops at the same site in 5- to 15-year jhum system, earthworms participated in N cycle through cast-egestion, mucus production and dead tissue decomposition. Soil N losses were more pronounced over a period of 15-year jhum system. The total soil N made available for plant uptake was higher than the total input of N to the soil through the addition of slashed vegetation, inorganic and organic manure, recycled crop residues and weeds.

10.5.5 Improvement Of Soil Physical, Chemical and Biological Properties

Limited studies on vermicompost indicate that it increases macropore space ranging from 50 to 500 m, resulting in improved air-water relationship in the soil which favorably affects plant growth [33]. The application of organic matter including vermicompost favorably affects soil pH, microbial population and soil enzyme activities. It also reduces the proportion of water-soluble chemical species, which cause possible environmental contamination.

10.5 VERMITECHNOLOGY IN WASTE MANAGEMENT

Open dumping is the prevailing method of solid waste disposal in many of the developing countries. It has become increasingly expensive and hazardous to the natural environment. Therefore, presently there is an urgent need to explore the potential of earthworms in waste management. For millennia, earthworms have been preparing soil for the colonization and evolution of plants. They have played a commendable role in agriculture. Their value in supporting the waste disposal and management systems is being realized by the day. Earthworms, in dense culture and in large quantities, can physically handle virtually any biological waste. Vermitechnology, based on this inherent ability of earthworms has the capacity to handle large quantities of organic wastes and is seen as a viable industrial process capable of sustained commercial operation (Sharma, S., et al, 2005).The core of the system is the process bed in which millions of worms are regularly fed with organic waste and from which worm stabilized organic matter, referred to as vermicompost, is harvested.

10.5.1 FEEDS FOR VERMITECH SYSTEMS

1. Animal Manures

Use of animal manure as primary feed for earthworms is very common in vermitech systems. Cattle solids are the most suitable of all animal wastes for earthworm biomass increase. They usually do not have materials that deter the growth of earthworms. Cowdung slurry is a suitable substrate for vermicomposting, both when mixed with solid materials or on applying to the surface of bedding materials containing earthworms. Horse manure is also suitable for the growth of earthworms. (Manaf, L.A., et al, 2009) Horse manure contains 0.7 % of nitrogen, 4.38 % of protein and 60 % of organic matter, trace amounts of phosphoric acid and potassium oxide (Ronald and Donald, 1977) and can therefore be applied directly as feed. Waste from the piggeries is probably regarded as the most productive refuse for growing the earthworms. Poultry wastes are higher in protein content, nitrogen and in terms of phosphoric acid than any other animal manure. The fresh waste generated from the poultry farms contains significant number of inorganic salts, and if used directly might threatened the survival of the worms (Edwards, 1988). These wastes have to be pretreated by composting, washing or simply by aging process to reduce the inorganic salt content and the heating potential.

2. Kitchen Waste and Urban Waste

The residential and the commercial portion make up to about 50 to 70 percent of the total Municipal solid wastes (MSW) generated in a community. The solid waste so generated can be of two types:

A. Biodegradable or organic- kitchen waste, straw, hay, paper and animal excreta.

B. Non-biodegradable- ash, stone, cinders, plastics, rubber and metals.

The food waste from domestic households and restaurants and other yard waste are good for growth of earthworms. Vegetable scraps from kitchen and other yard wastes provide ideal feed bed for growing earthworms. Earthworms digest municipal biosolids along with green mulch. About two thirds of this volume becomes vermicompost.

3. Paper Pulp and Card Board Solids

Paper and cardboard are excellent materials, both for feeding and for the bedding of earthworms due to their cellulose content. Earthworms convert cellulose into its food value faster than the proteins and other carbohydrates. These wastes do not need any special pretreatment and can be applied directly as a feed. In a recent work, Basheer, M and Agrawal,O.P.(2013) successfully utilized epigeic earthworm, Eudrilus eugeniae for the conversion of paper waste into vermicompost.

4. Compost and Waste Products

Spent mushroom compost is also a good medium to grow earthworms. It is low in plant nutrients. Brewery waste needs no modification, in terms of moisture and the worms can process it quickly.

5. Industrial Wastes

Wastes from the canning plant and potato chip or corn chip manufacturing unit are excellent food for worms. Wastes generated from vegetable oil factory are also suitable as feed (Kale, 1998). Wastes from logging and carpentry industries and sugar factories are also used as substrate to feed earthworms. When the earthworms are reared in the ratio of 1:1 sawdust and press mud, the cast generated shows 1.2 times more CFU (Colony Forming Units) than saw dust and 1.6 times more than the press mud (Parthasarathi,et al., 1999).Earthworms can partially detoxify wastes. The fly ash waste generated from the thermal power plants creates a major disposal problem due to its heavy metal content although it is supposed to be very rich in microbial biomass. It was found out that the organic waste, sisal green pulp, Parthenium and green grass cuttings admixed with 25% of fly-ash proved to be a potential valuable material for E.fetida biomass. The vermicompost so produced contains higher NPK content than the other available commercial manures. In some cases, earthworms are also used in the management of distillery waste containing wastes of malt, spent grain wash, yeast and molasses settled at the bottom of the lagoon. Seenappa, et al., (1995) observed that the total volume of cow dung leaf litter should be proportional to the total volume of distillery waste and press mud to have positive impact on the growth and production of worm biomass. Lakshmi and Vijayalakshmi (2000) reported that the filter press mud from the sugar factory could be used as a feed in the vermicomposting units. It is seen that after worms have worked on it, the press-mud is converted to nutrient rich manure and its physico-chemical features improved after vermicomposting. In 2011, Murali,M.,et al reported that coir wastes could be converted into vermicompost by utilizing Eudrilus Eugeniae. Studies by this author reported that chip wash residue, a kind of wood waste from wet-process hardboard factory could be converted into vermicompost by utilizing E. eugeniae (E.Sreenivasan, 2013a).

10.5.6 Other Applications

1. In aquaculture

In aquaculture, the most common method of solid waste disposal is land spreading, which causes pollution of soil, surface and ground water bodies resulting in untimely death of aquatic organisms. Vermicomposting of such waste controls water and soil pollution, thereby ensuring better survivability and growth offish, prawn and other aquatic organisms within its natural habitat. The application of vermi-castings, which is a high-grade organic fertilizer to the aquaculture ponds, reduces the input cost and makes the aquaculture process more profitable. It also helps in combating the harmful effects against chemical fertilizer if so, ever present in the receiving water. Large-scale vermiculture has the potential of supplying earthworm meal as a substitute for fishmeal. Earthworm meal contains all the essential amino acids required in fish feed. The methionine and lysine availability is higher than that of the normal fishmeal. There is also an increasing demand for protein-rich raw materials in other animalfeed industry.

2. Vermifilter Ground Water Recharge

Earthworm increases the hydraulic conductivity and natural aeration by granulating the clay particles. They also grind the silt and sand particles, increasing the total specific surface area, which enhances the ability to adsorb the organics and inorganics from the wastewater. This is ideal for dilute wastewater (such as sewage). Figure shows the vermifilter that could be used to recharge groundwater. The loading rate of wastewater is calculated as 2m2 /m3 of sewage. Earthworms ensure bio-sanitation and prevent sewage sickness through effective regeneration of adsorption ability with their bacterial farming along with their grazing act on surplus bacterial biomass (Bhawalkar, 1995).

Water Recovery

It is possible to have single or multiple stage vermifilter depending on the strength of the wastewater and desired quality of renovated water. In principle, a single unit can produce any given purity of water by increasing the recyclable ratio, which reduces the organic loading. The first stage achieves roughing filtration and the second stage achieves polishing (Bhawalkar, 1995).

11

TROUBLE SHOOTING

11.1 INTRODUCTION

A healthy worm bin should not have an odor. If you smell something bad, you are most likely overfeeding your worms. Too much food can deplete air flow and create an anaerobic environment. Take any undigested food out of the bin and replace it with new bedding.

11.1.1 Temperature

Heat as well as cold cause many problems for vermicomposting. A red wiggler becomes inactive once the temperature of the bedding rises above 29 °C. This could be avoided by placing the bin under shade at all times, if placed outdoors during the warm seasons. Evaporative cooling of the moist bedding keeps the bin cooler than the air temperature, but may need to add more water during the summer. The greater danger of overheating the worms arises from heat produced within the bin, which could be reduced substantially by feeding small amount of food frequently rather than a bulky food at one time. In general, worms like cool weather. They are at their highest activity and reproductive levels as the weather cools in the fall and warms in the spring.

11.1.2. Aeration

It is important to construct the bin to allow adequate airflow. Holes may be drilled on the upper sidewall of the bins for air circulation. Holes drilled on the lid may allow water inside during the rainy season. The type of bedding used also influences air circulation. Coarser bedding such as chopped leaves allows more air to circulate than fine textured bedding such as peat moss or shredded paper. As the composting process progresses, the bedding becomes more finely textured. This can be alleviated to some extent by periodically adding fresh bedding. Other ways to promote aeration includes occasional fluffing of the bedding material, avoidance of deep bedding (a maximum of 30cm), over-feeding and over-watering.

11.1.3. Acidity (pH)

The decomposition of organic matter produces organic acids that lower the pH of the bedding soil. The best way to deal with this is to add several cups of ground limestone to the bedding and in the application of Zeolite in proper amount. Limestone will serve dual purpose - maintaining the acidity and acting as a source of calcium to the worms. Other products, which can 40 be used, are powdered limestone, dolomite limestone. Baking soda should be avoided because of its high sodium content.

11.1.4. Pests and Diseases

Vermicompost worms are not subject to diseases caused by microorganisms, but they are subject to predation by certain animals and insects (red mites are the worst) and to a disease known as "sour crop" caused by environmental conditions. A brief overview of the most common pests (see Fig.11) and diseases is given below:

Ants. These tiny insects are a real problem because they not only consume the feed meant for the worms but are also found to attack young worms causing serious injuries. Ants are particularly attracted to sugar, so avoiding sweet feeds in the worm beds reduces this problem to a minor one. Or constructing a water- channel around the vermi-tank (at the bottom) will permanently solve the antproblem.

Rodents. Worms are a favorite natural food for many small animals like rats. So, if a rat gets access to your worm bed, you can lose a lot of worms very quickly. Rats and mice can chew through plastic or wood easily, and don't need very much space to move through a surface. This is usually only a problem when using openair systems in fields. It can be prevented by putting some form of barrier, by using wire mesh for the lids.

Birds. They are not usually a major problem, but if they discover your vermibeds, they will come around regularly for the worms. Putting a lid over the tank or cover over the material in the open-air systems, will eliminate this problem. These covers are also useful for retaining moisture and preventing too much leaching during rainfall events. Old carpet can be used for this purpose and is very effective.

Centipedes. These insects eat compost worms and their cocoons. Fortunately, they do not seem to multiply to a great extent within worm beds or windrows, so damage is usually light. If they do become a problem, one method suggested for reducing their numbers is to heavily wet (but not quite flood) the worm beds. The water forces centipedes and other insect pests (but not the worms) to the surface, where they can be destroyed by means of a hand-held propane torch or something similar.

Fig.1: Common pests of vermibeds

Sour crop. This is reported to be the result of poisoning caused by having too much protein in the bedding. This happens only when the worms are overfed. Protein builds up in the bedding and produces acids and gases as it decays. Farmers wishing to prevent sour crop should work by not overfeeding and by monitoring and adjusting pH on a regular basis. Keeping the pH at neutral or above will preclude the need for these measures.

Mite Pests. Insects are attracted to wormbeds due to its moist and organic environment. If the bedding is not properly maintained, acidity build up in the bedding soil may invite the mites as they are attracted towards an acidic, moist environment. Although small populations of mites thrive in all worm beds, they might create problems when present in excessive numbers. The mite populations at high level scan also cause worms to bury deep in their burrows without feeding.

11.1.5 White or Brown Mites

White or brown mites are not predaceous and tend to feed only on decaying or injured worms. However, during infestations, these mites can devour much of the food in earthworm beds, depriving earthworms from the nutrients.

11.1.6 Red Mites

These mites first appear as small white or gray clusters, resembling mould, which under magnification would reveal the clusters of juvenile red mites in various stages of development. The adult red mite is smaller than white or brown mites with bright red colour and an eggshaped body with four pair of legs. The red mites are known to be parasitic on earthworms. It attaches to the worm and relishes its coelomic fluid. They are also capable consuming the cytoplasmic fluids from egg capsules.

Mite Prevalence and Prevention

Harmful buildup of mites can be prevented by taking proper care of worm beds. The conditions usually associated with high mite population are:

a. Excess water: Beds that are too wet create conditions that are more favorable to mites than to the worms. Excessive wetting of beds may be avoided by adjusting watering schedules, improving drainage, and turning bedding frequently.

b. Over feeding: Excess food can lead to an accumulation of fermented feed in worm beds and lower the pH of the beds. The feeding schedules may be adjusted and modified according to seasonal variations. The pH of beds should be maintained to neutral (pH: 7), using calcium carbonate as the buffering agent

c. High moisture content feed or fleshy feed: vegetables with high moisture content can attract high mite populations in worm beds. Use of such feed should be limited, and if still, high mite populations persist, this feed should be discontinued until mite populations are under control.

Mite Removal

Several methods have been suggested for removing mites from earthworm beds. However, any type of mite removal, physical or chemical, will only be temporary unless worm-bed management is altered to make conditions less favorable for mites. The following techniques range from low- to high-intensity measures.

- The worm beds should be exposed to sunlight for several hours, however one should make sure that the earthworms are not directly exposed to sunlight. The amount of water and feed should be reduced. This will further encourage the mites to leave the beds.

- Moistened newspapers or burlap (jute) bags may be placed on top of the beds, and these can be removed as mites accumulate on them. This procedure may be repeated until mite populations are substantially reduced. * Pieces of watermelon or potato slices may be placed on top of the worm beds. The peels could then be removed with the mites.

- The bed may be watered heavily without flooding. This will compel the mites to move up to the surface. The mites can then be scorched using a hand-held propane torch. This procedure may be repeated several times, at three-day intervals, if needed.

- Light sulphur dusting will kill the mites. Or bed may be wetted (as suggested above) and then the sulphur added directly to the mites. Sulphur should be applied at the rate of approximately 2 g per 0.93 square meter of bed. Sulphur will not harm the worms, but in time, it may increase the acidity of the bed. In the past, some chemical pesticides have been used in worm beds. However, due to their biomagnification, it is not advisable to use these chemical compounds. Although safer miticides do exist in the market they are not specifically made for the Vermibed.

11.2 ODOUR PROBLEMS

The problem of unpleasant odours is caused by lack of oxygen in the compost due to overloading of food waste, and when the bin contents become wet. The solution is to stop adding food waste until the worms and microorganisms have broken down the initial feed and to gently stir up the entire contents to promote aeration. The drainage holes may be checked for blocking. If the drainage is insufficient additional holes can be drilled. Worms have been known to crawl out of the bedding if conditions are not favorable for them. If this migration is not triggered by moisture content of the soil, then the bedding may be acidic. Avoid adding citrus peels and other acidic foods to the bedding as these might reduce the pH of the bedding soil. One can overcome this acidic medium by adding a little garden lime and cutting down on acidic wastes.

In limiting the malodours, one should

-Reduce the amount of food

-Stir the bin thoroughly, especially at the bottom

-Add paper if the bedding is soggy

And if odours still persist, the best solution may be to start over, using new bedding, a minimal amount of scraps.

RELATED TERMINOLOGY

Aeration: Aeration the oxygen flow through a substance

Anaerobic: Anaerobic is the quality of lacking oxygen. Anaerobic vermicompost can be identified by a foul odor and is often densely compacted, thus preventing the medium from being exposed to sufficient oxygen levels.

Bedding: Worm bedding is a carbon-rich substrate in a worm bin that worms will inhabit. Worm bedding breaks down more slowly than nitrogen-rich food waste in a worm bin but can still serve as a food for the worms as it decomposes over time.

Bokashi: Bokashi is an anaerobic form of composting using inoculated bran to help ferment organic waste that may not be suitable for vermicomposting or traditional composting, such as meat and dairy products. Some vermicomposters claim success in feeding their worms bokashi.

Calciferous gland: A worm's calciferous gland excretes calcium carbonate (CaCO3) which scientists believe aids in digestion and reduces toxicity of metals and tannins. Calciferous gland activity can vary widely between worm species.

Clitellum: Clitellum known as the saddle of the worm, its the large pinky-orange band which creates slime to wrap sperm and eggs in after mating. It goes hard and slides off the worms, turning into a cocoon.

Clitellum: The clitellum is the the fleshy band on a worm nearer to the head and its presence indicates sexual maturity. The location, color, and texture of the clitellum are useful indicators or identifying a particular worm species.

Cocoon: The nutrient rich egg case which contains up to 20 embryonic worms, although only a few hatch

Cocoon: A worm cocoon is a small capsule, often the size of a pinhead or an apple seed, which contains approximately 3 embryonic worms. Cocoons take between 21 to 28 days to hatch juvenile worms and will change color as they mature.

coelomic Fluid : Coelomic fluid the slime which a worm produces to keep its skin moist and to protect it from predators

Compost: Compost is decomposed organic matter. Hot compost has undergone one if not several heating cycles where thermophilic microbes thrive and help to further break down the organic matter.

Composting Worm: Composting worm the type of worm which lives in topsoil and feeds on rotting vegetable waste

Crop: A sack behind the worms mouth where food it stored on its way down to the gizzard

Earthworker Worm: The type of worm which makes deep burrows in your garden and eats a diet of leaves and earth.

Eisenia Fetida: The binomial name of the Red Wiggler.

Eisenia Hortensis: The binomial name of the European Nightcrawler.

Epigeic: Epigeic worms live near the surface in loosely-packed leaf litter or detritus. Most composting worms are epigeic.

Ganglion : Ganglion a mass of neurons connected ultimately to the light sensors on the worms body

Gizzard : Gizzard a sack behind the crop, which is strong walled and acts as a food processor by mashing up the worm›s food with the help of grit.

Hatchlings : Hatchlings newly hatched worms, which are see through but perfectly formed with the right number of segments

Indian Blue Worm: Also known as perionyx excavatus, the Indian Blue Worm or "blue" is a composting worm that favors temperate to tropical climates. The Indian Blue is often confused with the red wiggler, but can be identified by its less prominent clitellum its noticeable elongation while moving.

Intestines: Where digestion and absorption of food takes ending up at the anus, where non-digested food and soil are excreted as worm castings

Landfill Purgatory: The place where old food and newspapers sit and fester for years and years giving off methane, a potent greenhouse gas.

Leachate: Leachate, often confused with worm tea, is excess moisture that has drained though a vermicomposting system. The liquid collected from the tap of a stackable plastic worm bin is leachate, not worm tea.

Microbes: Tiny living things, whether plant or animal; bacteria, protozoa, fungi, actinomycetes.

Oligochaeta: The class of segmented worms which earthworms belong to

Perionyx Excavatus: The binomial name of the Indian Blue Worm.

pH: A measure of the acidity or alkalinity of a substance. Measured from 14- 1, with 7 being neutral prostomium the tongue like pad above the worm›s mouth which gets it forge new burrows and get food into its mouth

Protein Poisoning: Protein poisoning happens when worms are fed a diet that is either acidic or high in protein, both of which will cause acidic conditions that lead to fermentation in the worm's gut. Having no way to counteract the gasses from the fermentation, the worm's intestines will rupture, resulting in the "string of pearls" appearance of a worm afflicted with protein poisoning.

Red Wigglers: Also known as *eisenia fetida*, Red Wigglers are the most common composting worm in North America and are suitable for a wide range of climactic conditions.

Root Dwelling Worm : Root dwelling worm the type of worm which in burrows lives deep under the surface and usually never comes above ground at any time.

Segment: A segment is the disc-shaped portion of a worm's body bound together from the anterior and posterior of the worm by a membrane.

Segments: Segments the ring like sections which worms are made from connected together by membranes.

Setae: Setae the bristle like hairs, found around the segments of worms, which help them move along

Species: Species a category of classification for any animals which can breed together and produce fertile offspring

Vermicast: Technically, vermicast is a single particle of worm castings, but the term is often used interchangeably with worm castings.

Vermicompost: Vermicompost is the byporoduct of vermicomposting or worm composting, normally existing of worm castings, humus, unprocessed organic materisal, and microorganisms.

Windrow: A windrow is a long pile of material, vermicompost in this case, normally heaped by a machine or manually. Vermicomposting in windrows is very common in warmer climates as it is not capital intensive.

Worm Castings: Worm castings are worm poop! They are the excretion of the earthworm after the digestion process. Worm castings are considered a highly-potent soil amendment that enriches soil and aids in plant growth, pest and pathogen suppression, water retention, soil aggregation, and other benefits.

Worm Casts: Worm casts worm poo- made of soil, undigested vegetable matter, some bacteria and lots of plant nutrients

Worm Tea: Worm tea, unlike leachate, is a deliberately-produced by-product of vermicompost made by suspending vermicompost in a fine mesh bag into a container of water which is typically aerated in order to boost the microbe populations within the tea. Applied as a soil drench or foliar spray, worm tea is considered a potent fertilizer.

Worm Bin: Like a compost bin, a worm bin is used as a repository for food scraps and other organic waste materials to be converted into organic fertilizer. As the name implies, a worm bin uses live worms to digest organic wastes in a process known as vermicomposting. Worm bins contain certain types of worm species in order to break down the waste material into organic soil conditioners.

Vermi Composting: The process whereby worms feed on slowly decomposing materials (e.g., vegetable scraps) in a controlled environment to produce nutrient-rich soil.

Vinyl Chloride: A chemical compound, used in producing some plastics, that is believed to be carcinogenic.

Volatile: Description of any substance that evaporates readily.

Volatile Organic Compound (voc): Any organic compound that participates in atmospheric photochemical reactions and has a high vapor pressure and low water solubility.

Waste:--Objects or materials for which no use or reuse are intended.

Waste-to-energy: The practice of processing waste products to generate steam, heat, or electricity. WASTE MANAGEMENT: The management of waste collection, handling, processing, storage and transport from where it is produced to where it is finally disposed.

Waste Prevention: An aspect of waste management that involves reducing the amount of waste we produce and minimizing the potential harm to human health or the environment from packaging or ingredients in products.

Waste recycling: A method of recovering waste to use them as resource materials. It involves the reuse of wastes or the collection and treatment of a waste product for use as a replacement of all or part of the raw material in the manufacturing process. It lessens new products and consumables that need to be produced, contributing to source reduction and reducing energy consumption.

Waste Reduction: Reduction in the quantity (mass) of material entering the waste stream as a result of a change in the goods, process or packaging.

Waste-to-Energy: The practice of processing waste products to generate steam, heat or electricity.

Waste Wood: It describes wood refuse that typically comes from packaging, construction and demolition. Domestic or household waste wood usually comes from old furniture, do-it-yourself construction and discarded wood materials from home renovations. It comprises a big chunk of annual waste. But instead of being disposed of in landfills, they can now be used as a source of biomass energy.

Water Cooling: It is an eco-friendly method used to lower the temperatures of computer processors, and sometimes other components such as graphics cards, using water rather than air as the cooling medium. Also known as "liquid cooling."

Water Footprint: An estimation of the amount of water used.

Wave Energy: Power derived from forces produced by ocean waves. Waves are caused by the wind blowing over the surface of the ocean.

Weee: Short for Waste Electrical and Electronic Equipment (WEEE), which are any unwanted devices with a plug or battery – from a remote control or digital camera to a vacuum cleaner or fridge freezer. These devices must be disposed of carefully to avoid damage to the environment. To get rid of an unwanted device, you can bring it to a civic amenity site or leave it with a retailer when you are buying a new device. All WEEE left in retail outlets and civic amenity sites are collected for recycling.

Wind Energy: Energy harnessed from the wind at wind farms and converted to power.

Wind Farm: A cluster of wind turbines used to harness power for producing electricity. Individual wind turbines are interconnected in a medium voltage power harnessing system, whose electrical current is boosted using a transformer. These farms do not consume fuel and do not emit pollution.

Wind Power: Wind power is the conversion of wind energy into a useful form of energy by using wind turbines to make electrical power, windmills for mechanical power, wind pumps for water pumping or drainage, or sails to propel ships.

Wind Turbine: It is a device that converts the kinetic energy of the wind into mechanical energy that can be used to drive equipment such as pumps. The addition of a generator allows the wind's kinetic energy to be converted into electricity. These are usually mounted on a tower.

Windrow Composting: It spreads organic materials into long, semi-circle shaped piles which are mechanically turned using heavy equipment to maintain even decomposition. Piles generally range from 4-8 feet in height and 14-16 feet in length. The organic matter in compost improves soil nutrient-holding and water-retaining capabilities. It reduces fertilizer requirements and erosion while enhancing soil tilth. It requires no source of electricity.

Wood Energy: Energy derived from wood in various forms.

Woodfree: Paper that does not contain mechanical wood pulp.

Woodfuel: It includes solids (fuel wood and charcoal), liquids (black liquor, methanol, and pyrolitic oil) and gases from the gasification of these fuels.

Wood Pellets: These are small particles used for energy generation made of dried, ground and pressed wood. Wood pellets are originally produced from wood waste (such as sawdust and shavings), rather than whole logs. The raw material is dried, mechanically fractioned to size, and extruded under intense pressure into pellets.

Wood Residues: Wood left behind in the forest after forest harvesting and wood by-products from wood processing, such as wood chips, slabs, edgings, sawdust and shavings.

Wood Waste: Unneeded or unusable wood from construction, demolition or renovation projects.

Xeriscaping: Refers to landscaping and gardening in ways that reduce or eliminate the need for supplemental irrigation. It is promoted in areas that do not have easily accessible supplies of fresh water, and is gaining acceptance in other areas as climate patterns shift.

Zero Emission: An engine, motor or other energy source that does not produce any gas or release any harmful gases directly into the environment.

Zero Emission: vehicles - Vehicles (usually powered by electricity) with no direct emissions from tailpipes or fuel evaporation.

Zero Energy Building (zeb): Buildings do not consume energy from conventional power plants and produce zero carbon emissions. A zero energy building does not rely on supply from the energy grid because it uses its own means to harvest energy on site, usually through solar, wind or hydro power. Some buildings that use very little electricity from the grid can be called near-zero energy buildings or ultra-low energy homes.

Zero Waste: Zero waste is a philosophy that encourages the redesign of resource life cycles so that all products are reused. A production system aims to eliminate the volume and toxicity of waste and materials by conserving or recovering all resources.

Ecosystem: A mutually dependent system consisting of plant, animal life and inorganic matter.

Fertilizer: A substance (natural or man-made) used to enrich the soil and to provide food for plants.

Food Scraps: In the Compost Module, food scraps generally refer to uncooked fruit and vegetable scraps or any compostable food materials.

Greens: Nitrogen rich compost materials (usually wet).

Heap: An unenclosed compost pile.

Humus: Finished compost, formed through the breakdown of plants and animal matter. Humus retains and slowly releases nutrients to plants.

Leachate: Liquid that has been generated by solid waste decomposition and which has extracted, dissolved or suspended materials in it. The leachate from a compost bin or worm bin is full of nutrients and is an excellent liquid fertilizer.

Leaf mold: Decomposed or mostly decomposed leaves.

Macroorganisms: Organisms that are visible to the eye.

Microorganisms: Organisms that cannot be seen without magnification.

Mulch: A layer of partially decomposed plant materials placed on top of garden beds and around plants and shrubs.

Organic matter: Any organic material that is or once was living or was once produced by a living organism.

Overload: To put too much food into a worm bin that can be processed aerobically.

Red Worm: A variety of earthworm suitable for vermicomposting. The Red Wiggler is a red worm.

Rodent Resistant: Compost bins designed or modified in such a way as to deter pests from making a home in the bin.

Screening: To sift out uncomposted matter from humus to create a fine compost.

Soil: Tiny rocks, sand, silt, clay plus decomposers plus organic matter.

Soil Conditioner: Something that enriches the physical condition of soil and increases its organic content.

Vermicompost: To carry out composting with worms or the end product from composting with worms. Vermicompost contains worm castings, broken down organic matter, bedding, worm cocoons, worms and other organisms.

Vermicomposter: A worm bin or person who composts with worms.

Vermicomposting: composting with worms.

Vermiculture: worm farming or raising earthworms.

Worm Bin: A container especially prepared for worms to live in and eat organic garbage. A vermicomposting system.

Worm Castings :worm manure or worm 'poop'

Wet Garbage: Usually refers to food scraps, grass clippings and garden waste; compostable, organic materials.

REFERENCE

K.I.M. Perera, A. Nanthakumaran, "Technical feasibility and effectiveness of vermicomposting at Household level Tropical Plant Research, pp. 51-57, 2015.

Londhe ,S. M. Bhosle, "Recycling of Solid Wastes in to Organic fertilizers using low cost treatment: Vermicomposting", International Journal of Innovations in engineering research and technology, pp. 9,2015.

P.Taylor , "Vermicomposting of Sewage Sludge: Earthworm Population and Agronomic Advantages Vermicomposting of Sewage Sludge: Earthworm Population and Agronomic Advantages", Compost Science & Utilization, pp. 37-41,2013.

R.Gupta, A. Yadav, V.K. Garg, "Influence of vermicompost application in potting media on growth and flowering of marigold crop," Int. J. Recycl. Org. Waste Agricult, pp. 47, 2014.

R.Thangaraj , "Leaf litter waste management by vermicomposting using local and exotic earthworm species", Journal of Science, pp. 314-319,2013.

S.Adhikary, "Vermicompost, the story of organic gold: a review", Agricultural Sciences, pp. 905-917,2012.

S.Paul, S.S.Bhattacharya, "Vermicomposted Water Hyacinth Enhances Growth and Yield of Marigold by Improving Nutrient Availability in Soils of North Bank Plain of Assam," Research & Reviews: Journal of Agricultural Science & Technology, pp. 36-46, 2012.

Sandeep, D. Singh, J.Yadav , Urmila, "Assessment of nutrient status of Vermicompost of leaf litter using Eisenia fetida," Journal of Entomology and Zoology Studies, pp.1135-1137,2017.

Sunil J. Kulkarni, "Vermicomposting- A Boon for Waste Minimization and Soil," International Journal of Research and Review- Vermicomposting, pp. 76-81, 2017.

K.Kamineni,P.Sidagam,"A Study On Recycling Organic Wastes Through Vermicomposting", International Journal Of Advanced Biotechnology And Research, Pp. 87,2014.

V.S.Varma, A.S.Kalamdhad , Khwairkpam, "Feasibility of Eudrilus Eugenia and Perionyx excavatus in vermicomposting of water hyacinth," Ecol. Eng, pp. 127-135,2016.

Yadav, R. Gupta, V. K. Garg, "Organic manure production from cow dung and biogas plant slurry by vermicomposting under field conditions," International Journal of Recycling of Organic Waste, pp. 1, 2013.

Abafita R., Shimbir, T. & Kebede T. 2014. Effects of Different Rates of Vermicompost as Potting Media on Growth and Yield of Tomato (Solanum lycopersicum L.) and Soil Fertility Enhancement. Sky Journal of Soil Science and Environmental Management, 3(7), 73-77.

Adiloğlu S., Eryılmaz Açıkgöz F., Solmaz Y. & Çaktü E. 2018, Effect of Vermicompost on the Growth and Yield of Lettuce Plant (Lactuca sativa L. var. crispa). International Journal of Plant & Soil Science,, 21(1), 1-5.

Ahirwar C. S. & Hussain A. 2015. Effect of Vermicompost on Growth, Yield and Quality of Vegetable Crops . International Journal of Applied and Pure Science and Agriculture, 1(8), 49-56.

Aksu G., Köksal S. B. & Altay H. 2017. Vermikompostun Bazı Toprak Özellikleri ve Pazı Bitkisinde Verim Üzerine Etkisi. ÇOMÜ Ziraat Fakültesi Dergisi, 5 (2), 123-128.

Alam M. N., Jahan M. S., Ali M. K. & Ashraf M. K. 2007. Effect of Vermicompost and Chemical Fertilizers on Growth, Yield and Yield Compenents of Potato in Barind Soils of Bangladesh, Journal of Applied Sciences Research,, 3(12), 1879-1888.

Arancon N. Q., Edwards C. A. & Lee S. S. 2007. Management of Plant Parasitic Nematode Populations by Use of Vermicomposts. 1735 Neil Avenue, Columbus, USA, 705-710 ref.11.

Arancon N. Q., Edwards C. A., Bierman P., Wlch C. & Metzger J. D. 2004. Influences of Vermicomposts on Field Strawberries. 1. Effects on Growth and Yields. Bioresource Technology, (93), 145 - 153.

Azarmi R., Giglou M. T. & Taleshmikail R. D. 2008. Influence of vermicompost on soil chemical and physical properties in tomato (Lycopersicum esculentum) field. African Journal of Biotechnology, 7(14).

Bai B. A. & Malakout M. J. 2007. The effect of different organic manures on some yield and yield quality parameters in onion. Iran Soil and Water Sciences Journal, 1(21), 33- 43.

Baker G. H. 1994. Earthworm, new discoveries, Rural Research. A CSIRO Quarterly, 163. 19– 23.

Buckerfield J. & Webster K. A. 1998. Worm Worked Waste Boosts Grape Yields Prospects for Vermicompostuse in Vineyards. Australia and New Zealand Wine Industry Journal, 13, 73-76.

Cihangir H. & Öktem A. 2015. Diyarbakır Koşullarında Farklı Organik Bitki Besleme Uygulamalarının Tatlı Mısır Bitkisinin (Zea mays L. saccharata Sturt) Taze Koçan Verimi Üzerine Etkisi. UÜ. Ziraat Fakültesi Dergisi, 29(2), 69-81.

Çıtak S., Sönmez S., Koçak F. & Yasin S. 2011. Vermikompost ve Ahır Gübresi Uygulamalarının Ispanak Bitkisinin Gelişimi ve Toprak Verimliliği Üzerine Etkileri. Derim, 28(1), 56-69 ref:31.

Demir H., Polat E. & Sönmez İ. 2010. Ülkemiz İçin Yeni Bir Organik Gübre: Solucan Gübresi. Tarım Aktüel, (14), 54 - 60. Deveci M. & Şalk A. 1995. Tekirdağ Şartlarında Ispanak Yetiştiriciliğinde Farklı Ekim Zamanı ve Bitki Sıklığının Gelişme ve Verim Üzerine Etkisi. Tekirdağ Ziraat Fak. Dergisi, 4(1-2), 1-11.

Edwards C. A. & Arancon N. Q. 2004. The use of earthworms in the breakdown oforganic wastes to produce vermicomposts and animal feed protein. In Earthworm Ecology. London, New York, Washington: 2nd Edition, Editor C.A. Edwards. C.R.C. Press, Boca Raton, Fl.,. Edwards C. A. & Bohlen P. J. 1995.

Biology and Ecology of Earthworms. (3 b.). New York,: Springer Hollanda. doi:978-0-412-56160-3 Edwards C. A. 1995. Commercial and environmental potential of vermicomposting: A historical overview. BioCycle, June, 62-63.

Entry J. A., Wood B. H., Edwards J. H. & Wood C. W. 1997. Influence of organic byproducts and nitrogen source on chemical and microbiological status of an agricultural soil. Biology Fertilizer Soil, 24, 196-204.

Ferreira L. D. & Merchant K. A. 1992. Field Research in Management Accounting and Control: A Review and Evaluation. Accounting, Auditing & Accountability Journal, Volume 5, Number 4, 0:0.

Fosgate O. T. & Babb M. R. 1972. Biodegradation of Animal Wastes by Lumbricus Terrestris. J. Dairy Sci, 55, 870-872.

Gutiérrez-Miceli F.A. Santiago-Bórraz J., Montes Molina J. A., Nafate C. C., Abdudarchila, M., Oliva LIven M.A. & Deendoven, L. 2007. Vermicompost as a Soil Supplement to Improve Growth, Yield and Fruit Quality of Tomato (Lycopersicum esculentum). Bioresource Technology, 98, 2781-2786. .

Jackson M. L. 1967. Soil Chemical Analysis. Yeni Delhi: Prentice Hall of India Pvt. Ltd. 8 .30, 2021 tarihinde https://books.google.com.tr/books?hl=tr&lr=&id= false adresinden alındı

Jahan F. N., Shahjalal A. M., Paul A. K. & Mehraj H. 2014. Efficacy of vermicompost and conventional compost on growth and yield of cauliflower. Bangladesh Research Publications Journal, 10(1), 33-38.

Kale R. D., Bano K., Vinayak K. & Bagyaraj D. J. 2013. Suitability Of Neem Cake As An Additive In Earthworm Feed And Its Influence On The Establishment Of Microflora. Journal of Soil Biology & Ecology, 6 (2), 98-103.

Kızılkaya R., Turkay F. H., Turkmen C. & Durmuş, K. 2012. Vermicompost Effects on Wheat Yield and Nutrient Contents in Soil Plant. Archieves of Agronomy and Soil Science, 58, 175-179.

Köksal S. B., Aksu G. & Altay H. 2017. Vermikompostun Bazı Toprak Özellikleri ve Pazı Bitkisinde Verim Üzerine Etkisi. ÇOMÜ Zir. Fak. Derg. (COMU J. Agric. Fac.), 5 (2), 123–128.

Kumar A. A. & Gupta R. K. 2018. The Effects of Vermicompost on Growth and Yield Parametres of Vegetable Crop Radish (Raphanus sativus). Journal of Pharmacognosy and Phytochemistry,, 7(2), 589-592.

Küçükyumuk Z., Gültekin M. & Erdal G. 2014. Vermikompost ve Mikorizanın Biber Bitkisinin Gelişimi ile Mineral Beslenmesi Üzerine Etkisi. Süleyman Demirel Üniversitesi Ziraat Fakültesi Dergisi,, 9 (1), 51-58.

Leon A. P., Martin J. P. & Chiesa A. 2012. Vermicompost Application and Growth Patterns of Lettuce (Lactuca sativa L.). Agricultura Tropica et Subtropica, 45(3), 134-139.

Maji D., Misra P., Singh S. & Kalra A. 2017. Humic Acid Rich Vermicompost Promotes Plant Growth by Improving Microbial Community Structure of Soil as Well as Root Nodulation and Mycorrhizal Colonization in the Roots of Pisum sativum. Applied Soil Ecology , 110, 97-108.

Maltaş A. Ş., Tavalı İ. E., Uz İ. & Kaplan M. 2017. Kırmızı Baş Lahana (Brassica oleracea var. capitata f. rubra) Yetiştiriciliğinde Vermikompost Uygulaması. Akdeniz Üniversitesi Ziraat Fakültesi Dergisi., 30(2), 155-161.

Mısırlıoğlu M. 2011. Toprak Solucanları Biyolojileri, Ekolojileri ve Türkiye Türleri,. Ankara: Nobel Akademik Yayıncılık. Müftüoğlu N. M. 2016. Vermikompostun Ispanak Verimi ve Bazı Toprak Özellikleri Üzerine Etkisi. ÇOMÜ Ziraat Fakültesi Dergisi (COMU J. Agric.Fac.), 4(1), 1-5.

Nagavallemma K. P., Wani S. P., Lacroix S., Padja V. V., Vineela C., Rao B. & Sahrawat, K. L. 2006. Vermicomposting: Recycling Wastes into Valuable Organic Fertilizer. Global Theme on Agroecosystems Report no. 8, 2(1), 1-16.

Nurhidayati N., Ali U. & Murwani I. 2016. Yield and Quality of Cabbage (Brassica oleracea L. var. capitata) under Organic Growing Media Using Vermicompostand Earthworm Pontoscolex corethrurus Inoculation. Agriculture and Agricultural Science Procedia, 11, 5-13.

Özkan N. & Müftüoğlu N. M. 2015. Farklı Dozlardaki Vermikompostun Marul Verimi Ve Bazı Toprak Özellikleri Üzerine Etkisi. Atatürk Bahçe Kültürleri Araştırma Enstitüsü Dergisi., 45 (Özel sayı), Cilt:2, 121-124. 2021 tarihinde alındı

Pant P. A., Theodore J. R., Hue N. V., Talcott S. T. & Krene K. 2009. Vermicompost Extracts Influence Growth, Mineral Nutrients, Phytonutrients and Antioxidant Activity in

Pak Choi (Brassica rapa cv. Bonsai, Chiensis group) Grown Under Vermicompost and Chemical Fertiliser. Journal of The Science of Food and Agriculture, 89, 2383-2392.

Pascual J. A., Ayuso M., Hernández T. & García C. 1997. Phytotoxicity and fertilizer value of different organic materials. Agrochemical, 41, 50-62. Peyvast G. H., Olfati J. A., Madeni S. & Forghani A. 2007. Effect of Vermicompost on the Growth and Yield of Spinach (Spinacia oleracea L.). J. of Food, Agric. & Environ, 6(1), 132-135.

Pritam. S., Garg V. K. & Kaushik C. P. 2010. Growth and Yield Response of Marigold to Potting Media Containing Vermicompost Produced from Different Wastes. Environmentalist, 30, 123-130.

Rivasol 2021. Solucan Gübresinin Bilinmesi Gereken Faydaları. rivasol.com.tr: https://www.rivasol.com.tr/solucan-gubresinin-bilinmesi-gereken-faydalari adresinden alındı

Roberts P., Ljones D. & Edward-Jones G. 2007. Yield and Vitamin C Content of Tomatoes Grown in Vermicomposted Wastes. Journal Science Food Agriculture, 87, 1957-1963. Sağlam N., Doksöz S., Geboloğlu N., Şahin S. & Yılmaz E. 2015. Agrimol Örtü ve Sıvı Solucan Gübresinin Farklı Uygulama Sayısı ve Dozlarının Kıvırcık Yapraklı Salatada Verim, Kalite ve Bitki Gelişimine Etkileri. Tarım Bilimleri Araştırma Dergisi, 8(1), 59- 61. Shirkhodaei M., Darzi M. T. & Hadi M. H. 2014. Influence of Vermicompost and Biostimulant on The growth and Biomass of Coriander (Coriandrum sativum L.). International Journal of Advanced Biological and Biomedical Research, 2(3), 706 - 714. Singh R., Sharma R. R., Kumar S., Gupta R. K. & Patil R. T. 2008. Vermicompost Substitution Influences Growth, Physiological Disorders, Fruit Yield and Quality of Strawberry (Fragaria x ananassa Duch). Bioresource Technology, 99, 8507-8511. Sinha R. K., Herat S., Dalsukh V. & Kurunal C. 2009. The Concept of Sustainable Agriculture: An Issue of Food Safety and Security for People, Economic Prosperity for the Farmers and Ecological Security for the Nations. American-Eurasian J. Agric & Environ. Sci, 5 (S), 01-5.

Sönmez İ., Sönmez S. & Kaplan M. 2002. Çöp kompostunun bitki besim maddesi içerikleri ve bazı organik gübrelerle karşılaştırılması. Selçuk Üniversitesi Ziraat Fakültesi Dergisi, 16(29), 31-38.

Suthar S. 2009. Impact of Vermicompost and Composted Farmyard Manure on Growth and Yield of Garlic (Allium sativum L.) Field Crop. International Journal of Plant Production, 3(1), 27-38.

Suthar S. 2010. Evidence of Plant Hormone Like Substances in Vermiwash: An Ecologically Safe Option of Synthetic Chemicals for Sustainable Farming. Ecological Engineering, 36, 1089-1092. Şalk A. 1992.

Özel Sebzecilik 1. Ders Notları Tekirdağ Üniversitesi Ziraat Fakültesi. Özel Sebzecilik 1. Ders Notları (Basılmamış) Tekirdağ. Tavalı İ. E., Maltaş A. Ş., Uz İ. & Kaplan M. 2013.

Karnabaharın (Brassica oleracea var. botrytis) Verim, Kalite ve Mineral Beslenme Durumu Üzerine Vermikompostun Etkisi. Akdeniz Üniv. Ziraat Fakültesi Dergisi, 26(2), 115-120.

Tavalı İ. E., Maltaş A. Ş., Uz İ. & Kaplan M. 2014. Vermikompostun Beyaz Baş Lahananın (Brassica oleracea var. Alba) Verim, Kalite ve Mineral Beslenme Durumu Üzerine Etkisi. Akdeniz Üniversitesi Ziraat Fakültesi Dergisi, 27(1), 61-67.

Tomati U. & Gali E. 1995. Earthworms, soil fertility and plant productivity. Acta Zoologica Fennica, 196, 11-14.

Ulusu F. & Yavuzaslanoğlu E. 2017. Örtü Altı Organik Domates Yetiştiriciliğinde Farklı Gübre Uygulamalarının Bitki Yeşil Aksamı ve Meyve Verimine Etkisi. Türk TarımGıda Bilim ve Teknoloji Dergisi, 5(13), 1757-1761.

Venkatesh P. B., Patil C. V. & Giraddi R. S. 1998. Effect of In situ Vermiculture and Vermicompost on Availability and Plant Concentration of Major Nutrients in Grapes. Karnataka Journal Agricultural Science , 11(1), 117 - 121.

Wang C., Sun Z. J. & Zheng D. 2006. Research advance in antibacterial immunity ecology of earthworm. The Journal of Applied Ecology, 17(3), 525.

Yağmur B. & Eşiyok D. 2019. solucan-gubresi-vermikompost-iiivermikompostun-kullanimalanlari. 8 30, 2021 tarihinde www.dunyagida.com.tr: http://www.dunyagida.com.tr/haber/solucan-gubresi-vermikompostiiivermikompostun-kullanim-alanlari/4341

Yourtchi M. S., Hadi M. & Darzi M. T. 2013. Effect of Nitrogen Fertilizer and Vermicompost on Vegetative Growth, Yield and NPK Uptake by Tuber of Potato. International Journal of Agriculture and Crop Sciences, 5(18), 2033-2040.

Zimny L., Malak D. & Sniady R. 2001. Yielding of Sugar Beet Cultivated After Manure and Vermicompost in the Background of Increasing Doses of Nitrogen Fertilization. Archives of Agronomy and Soil Science, (47), 473-480.

Khaleel R, Reddy KR, Overcash MR (1981) Changes in soil physical properties due to organic waste applications: A review. J Environ Qual 10: 133-141. Link: https://goo.gl/5363pR

Butler TA, Sikor LPM, Steinhilber L, Douglass LW (2001) Compost Age and Sample Storage Effects on Maturity Indicators of Biosolids Compost. J Environ Qual 30: 2141-2148. Link: https://goo.gl/xBMrJa

Senesi N, Plaza C, Brunetti G, Polo A (2007) A comparative survey of recent results on humic-like fractions in organic amendments and effects on native soil humic substances. Soil Biol Biochem 39: 1244-1262. Link: https://goo.gl/WvoABr 4. Kumar KU, Henock, Tsegay (2015) Conversion of solid waste into bio fertilizer by vermicomposting a case study of padmanadapuram. International Journal of Innovative Research in Science, Engineering and Technology.

3801- 3808. Link: https://goo.gl/fJK07Z 5. Sequeira V, Chandrashekar JS (2015) Vermicomposting of biodegradable municipal solid waste using indigenous Eudrilus sp. Earthworms. Int J Curr Microbiol App Sci 4: 356-365. Link: https://goo.gl/nzr1am

Aali R, Jafarpour M, Kazemi E, Pessarakli M (2017) Effects of raw materials on vermicompost qualities. Journal of Plant Nutrition. Link: https://goo.gl/lSfzg0

Julka JM (1986) Earthworms resources of India Proc. Nat. Sem. Org. waste utilization, Vermicompt, Part B: verms and Vermicomposting. In: Dash RC, Senapathi BK, Mishra PC, (Eds.) 1-7.

Pathma J, Sakthivel N (2012) Microbial diversity of vermicompost bacteria that exhibit useful agricultural traits and waste management potential. Springer Plus 1: 26. Link: https://goo.gl/pv6t39

Edwards CA, Dominguez J, Neuhauser EF (1998) Growth and reproduction of Perionyxexcavatus (Perr.) (Megascolecidae) as factors in organic waste management. Biology and Fertility of Soils 27: 155-161. Link: https://goo.gl/gJ7OOO

Atiyeh RM, Lee S, Edwards CA, Arancon NQ, Metzger JD (2002) The infl uence of humic acids derived from earthworm-processed organic wastes on plant growth. Bioresource Technology 84: 7-14. Link: https://goo.gl/x3ixfi

Nagavallemma KP, Wani SP, Lacroix S, Padmaja VV, Vineela C, et al. (2004) Vermicomposting: Recycling wastes into valuable organic fertilizer. Global Theme on Agroecosystems Report no. 8. Patancheru 502 324, Andhra Pradesh, India: International Crops Research Institute for the Semi-Arid Tropics 20: 75-83. Link: https://goo.gl/xvD7sz

Ramesh P, Singh M, Rao AS (2005) Organic farming: it's relevant to Indian context. Current Science 88: 561-568. Link: https://goo.gl/CjZKP0

Westerman PW, Bicudo JR (2005) Management considerations for organic waste use in agriculture. Bioresource Technology 96: 215-221. Link: https://goo.gl/mZA0ut

Nair J, Sekiozoic V, Anda M (2006) EVect of pre-composting on vermicomposting of kitchen waste. Bioresour Technol 97: 2091–2095. Link: https://goo.gl/OE5Yzf

Champar-Ngam N, Iwai CB, Ta-Oun M (2010) Vermicompost: tool for agroindustrial waste management and sustainable agriculture. International Journal of Environment and Rural Development 1-2: 38-43. Link: https://goo.gl/FmGxeT

Shamini K, Fauziah SH, Emenike CU (2011) Vermicomposing of spent tea: A sustainable approach for solid waste management. Proceeding of the 12th International conference on Environmental Science and Technology, Rhodes, Greece, 8-10 September A-1695-A-1700. Link: https://goo.gl/74glhZ

Khalid A, Arshad M, Anjum M, Mahmood T, Dawson L (2011) The anaerobic digestion of solid organic waste. Waste Management 31: 1737-1744. Link: https://goo.gl/78rmje

Lim SL, Wu TY, Lim PN, Shak LPY (2014) The use of vermicompost in organic farming: overview, effects on soil and economics. J Sci Food Agric 95: 1143- 1156. Link: https://goo.gl/E2tdDb

Adhikary S (2012) Vermicompost, the story of organic gold: A review. Agricultural Sciences 3: 905-917. Link: https://goo.gl/pZ9fxa

Dhimal M, Gautam I, Tuladhar R (2013) Effectiveness of vermicomposting in management of organic wastes using Eisenia foetida and Perionyx favatus in central Zoo Jawalakhel, Nepal. J Nat Hist Mus 27: 92-106. Link: https://goo.gl/G3X7PR

Mehta N, Karnwal A (2013) Solid waste management with the help of vermicomposting and its applications in crop improvement. Journal of Biology and Earth Sciences 3: B8-B16. Link: https://goo.gl/dXQIxr

Kumari S (2013) Solid waste management by vermicomposting. International Journal of Scientifi c & Engineering Research 4: 1-5. Link: https://goo.gl/pwSr2X

Nasiru A, Ismail N, Ibrahim MH. (2013) Vermicomposting: Tool for sustainable ruminant manure management. Journal of Waste Management 2013: 7. Link: https://goo.gl/bo2GKL

Kaouachi A, Ibijbijen J, Amane M, Jaafari SE (2013) Management of olive mill waste employing vermicomposting technology. International Journal of Science and Research 4: 486-490. Link: https://goo.gl/Me6Ovs

Pirsaheb M, Khosravi T, Sharafi K. (2013) Domestic scale vermicomposting for solid waste management. International Journal of Recycling of Organic Waste in Agriculture 2: 4. Link: https://goo.gl/JFOIZk

Sumi MG, Vani M, Idicula DV, Mini KD (2014) Solid waste management using vermicomposting and kodmic? bio-pedestal column and its utility as organic manure. Asian Journal of Microbiology, Biotechnology & Environmental Sciences 16: 333-338. Link: https://goo.gl/A0NulQ

Sadasivuni S, Bhat R, Pallem C (2015) Recycling potential of organic wastes of arecanut and cocoa in India: a short review. Environmental Technology 4: 91-102. Link: https://goo.gl/CMjcLm

Hasheminajs K, Kalbasi M, Golchin A, Shariatmadari H (2006) Comparison of vermicompost and composts as potting media for growth of tomatoes. Journal of Plant Nutrition 27: 1107-1123. Link: https://goo.gl/llEnPq

Lokeshwari M, Swamy CN (2008) Vermicomposting of municipal and agricultural solid waste with sewage sludge. Journal of Environmental Research and Development 3: 51-61. Link: https://goo.gl/0TnU4b

Suthar S (2009) Bioremediation of agricultural waste through vermicomposting. Bioremediation Journal 13: 21-28. Link: https://goo.gl/2knSBg

Gurav MV, Pathade GR (2011) Production of vermicompost from temple waste (Nirmalya): a case study. Universal Journal of Environmental Research and Technology 1: 182-192. Link: https://goo.gl/skuFBu

Beohar PA, Srivastava RK (2011) Poultry waste management through vermicomposting employing exotic and indigenous species of earthworms. Journal of Soil Science 1: 4-11. Link: https://goo.gl/R8kBhf

Lim PN, Wu TY, Sim EYS, Lim SL (2011) The potential reuse of soybean husk as feedstock of Eudrilus eugeniae in vermicomposting. Journal of the Science of Food and Agriculture 91: 2637-2642. Link: https://goo.gl/24YpoG

Dhimal M, Gautam I, Tuladhar R (2013) Effectiveness of vermicomposting in management of organic wastes using Eisenia foetida and Perionyx favatus in central Zoo Jawalakhel, Nepal. J Nat Hist Mus 27: 92-106. Link: https://goo.gl/oO7Ltp

Kaouachi A, Ibijbijen J, Amane M, Jaafari SE (2013) Management of olive mill waste employing vermicomposting technology. International Journal of Science and Research 4: 486-490. Link: https://goo.gl/TzCNuw

Londhe PB, Bhosale SM (2015) Recycling of solid wastes into organic fertilizers using low cost treatment: vermicomposting. International Journal of Innovations In Engineering Research and Technology 2: 1-11. Link: https://goo.gl/0FouGw

Nag AK, Singh B, Singh KK (2015) A pilot scale solid waste management programme through vermicomposting of organic waste worship materials from some religious places of Patna Bihar. Indian Journal of Applied Research 5: 297-301. Link: https://goo.gl/CiERlW

Albasha MO, Gupta P, Ramteke PW (2015) Management of kitchen waste by vermicomposting using earthworm, Eudrilus eugeniae. International Conference on Advances in Agricultural, Biological & Environmental Sciences (AABES-2015) J22-23. Link: https://goo.gl/QHcTI5

Jain N (2016) Waste management of temple fl oral offerings by vermicomposting and its effect on soil and plant growth. International Journal of Environmental & Agriculture Research 2: 89-94. Link: https://goo.gl/loRasI

Peyvast GH, Olfati JA, Madeni S, Forghani A, Samizadeh H (2008) Vermicompost as a soil supplement to improve growth and yield of parsley. International Journal of Vegetable Science 2: 19-27. Link: https://goo.gl/2gDG9K

Kizilkaya R, Turkay FSH, Turkmen C, Durmus M (2012) Vermicompost effects on wheat yield and nutrient contents in soil and plant. Archives of Agronomy and Soil Science 58: S175-S179. Link: https://goo.gl/95MnYP

Ansari A, Hanief A (2015) Microbial degradation of organic waste through vermicomposting. International Journal of Sustainable Agricultural Research 2: 45-54. Link: https://goo.gl/uedSi4

Karmakar S, Adhikary M, Gangopadhyay A, Brahmachari K (2015) Impact of vermicomposting in agricultural waste management vis à vis soil health care. J Environ Sci Natural Resources 8: 99-104. Link: https://goo.gl/akoWXX

Bharti N, Barnawal D, Shukla S, Tewari SK, Katiyar RS (2016) Integrated application of Exiguobacterium oxidotolerans, Glomus fasciculatum, and vermicompost improves growth, yield and quality of Mentha arvensis in salt-stressed soils. Industrial Crops and Products 83: 717-728. Link: https://goo.gl/A6Otth

Xu L, Yan D, Ren X, Wei Y, Zhou J, et al. (2016) Vermicompost improves the physiological and biochemical responses of blessed thistle (Silybum marianum Gaertn.) and peppermint (Mentha haplocalyx Briq) to salinity stress. Industrial Crops and Products 94: 574-585. Link: https://goo.gl/w7SsyX

Zamani J, Afyuni M, Sepehrnia N, Schulin R (2016) Opposite effects of two organic waste on the physical quality of an agricultural soil. Archives of Agronomy and Soil Science 62: 413-427. Link: https://goo.gl/eadfXG

Masullo A (2017) Organic wastes management in a circular economy approach: rebuilding the link between urban and rural areas. Ecological Engineering 101: 84-90. Link: https://goo.gl/p1OrYc

Malińska K, Golanska M, Caceres R, Rorat A, Weisser P, et al. (2017) Biochar amendment for integrated composting and vermicomposting of sewage sludge – The effect of biochar on the activity of Eisenia fetida and the obtained vermicompost. Bioresource Technology 225: 206-214. Link: https://goo.gl/e7Nx91

Maji D, Misra P, Singh S, Kalra A (2017) Humic acid rich vermicompost promotes plant growth by improving microbial community structure of soil as well as root nodulation and mycorrhizal colonization in the roots of Pisum sativum. Applied Soil Ecology 110: 97-108. Link: https://goo.gl/WHGpNr

Bhat SA, Singh J, Vig AP (2017) Earthworms as organic waste managers and biofertilizer producers. Waste and Biomass Valorization 1-15. Link: https://goo.gl/JSZG0y

Ansari AA, Ismail SA. A case study on organic farming in Uttar Pradesh. J Soil Biol. Ecol. 2001; 27:25-27.

Ansari AA, Sukhraj K. Effect of vermiwash and vermicompost on soil parameters and productivity of okra (Abelmoschus esculentus) in Guyana. African Journal of Agricultural Research. 2010; 5(14):1794-1798.

Arancon NQ, Edwards CA, Bierman P. Influences of vermin composts on field strawberries: Part 2. Effect on soil microbiological and chemical properties. Bioresource Technology. 2006; 97:831-840.

Balam SB. Studies on bio pesticidal effect of vermiwash in control of some foliar pathogens. M.Sc. (Ag) thesis submitted to Dr. Balasaheb Sewant Konkan Krishi Vidyapeeth, Dapoli, Dist. Ratnagiri, Maharashtra, 2000.

Bucker field JC, Flavel T, Lee KE, Web Ster KA. Vermicompost soil and liquid form as plant growth promoter. Pedobio logia. 1999; 42:753-759.

Chaoui HI, Zibilske LM, Ohno T. Effects of earthworm casts and compost on soil microbial activity and plant nutrient availability. Soil Biology and Biochemistry. 2003; 35:295-302.

Chattopadhyay A. Effect of vermiwash of Eisenia foetida produced by different methods on seed germination of green mung, Vigna radiate. International Journal of Recycling of Organic Waste in Agriculture. 2015; 4(4):233-237.

Edwards CA, Domínguez J, Arancon NQ. The influence of vermin composts on plant growth and pest incidence. In, S.H Shakir and W.Z.A. Mikhaïl, (Eds. Soil Zoology for Sustainable Development in the 21st century, 2004, 397-420.

Esakkiammal B, Lakshmibai L, Sornalatha S. Studies on the combined effect of vermicompost and vermiwash prepared from organic wastes by earthworms on the growth and yield parameters of dolichous lab Lab. Asian Journal of Pharmaceutical Science and Technology. 2015; 5(4):246-252.

Fathima M, Sekar M. Studies on Growth Promoting effects of Vermiwash on the Germination of Vegetable Crops. International Journal of Current Microbiology and Applied Sciences. 2014; 3(6):564-570.

Nath G, Singh K, Singh DK. Chemical Analysis of Vermi composts/Vermiwash of Different Combinations of Animal, Agro and Kitchen Wastes. Australian Journal of Basic Applied Sciences. 2009; 3(4):3671-3676.

Hatti SS, Londonkar RL, Patil SB, Gangawane AK, Patil CS. Effect of Perionyx excavates vermiwash on the growth of plants. Journal of crop Science. 2010; 1(1):1-5.

Hidlago PR, Matta FB, Harkess RL. Physical and Chemical properties of substrates containing earthworm castings and effects on marigold growth. Hortscience. 2006; 41:1474-1476.

Jayabhaye MM, Bhalerao SA. Influence of Vermiwash on Germination and Growth Parameters of Seedlings of Green gram (Vigna radiata L.) and Black gram (Vigna mungo L.). International Journal of Current Microbiology and Applied Sciences. 2015; 4(9):635-643.

Jayashree MP. Vermiwash - The wonder tonic in agriculture. Kissan World. 2006; 6:44.

Kale KE. Earthworm: Cinderella of Organic farming, Bangalore: Prism books Pvt. ltd. 1998, 88.

Kale RD. Earthworms Nature⊠s gift for utilization of organic wastes. In earthworm⊠s ecology. Edwards, C.A (Ed.) CRC Press LLC. BOCCA. Raton, Florida, 1998, 355-376.

Karuna K, Patil CR, Narayanswamy P, Kale RD. Stimulatory effect of earthworm body fluid (Vermiwash) on crinkle red varity of Anthurium andreanum Lind. Crop Research. 1999; 17(2):253-257.

Kumar P, Shekhar C, Basoli M, Kumar V. Sequential spray of vermiwash at critical stages influences growth and quality in gladiolus cv. white prosperity. Annals of Horticulture. 2013; 6(1):71-75.

Lalitha R, Fathima K, Ismail SA. The impact of biopesticide and microbial fertilizers on productivity and growth of Abelmoschus esculentus. Vasundara the Earth. 2000; (1-2):4-9.

Gopal M, Gupta A, Palaniswami C, Dhanapal R, Thomas GV. Coconut leaf vermiwash: a bio-liquid from coconut leaf vermicompost for improving the crop production capacities. Current Science. 2010; 98:1202-1210.

Mishra K, Singh K, Tripathi CPM. Journal homepage: http://www. Journal. Ijar. Com. International Journal of Advanced Research. 2014; 2(1):780-789.

Mishra K, Singh K, Tripathi CPM. Organic farming of rice crop and management of infestation of Leptocorisa varicornis through combined effect of vermiwash with biopesticides. Research Journal of Science and Technology. 2015; 7(4):205-211.

Nath G, Singh K. Effect of vermiwash of different vermin composts on the kharif crops. Journal of Central European Agriculture. 2012; 13(2):379-402.

Nath G, Singh K. Combined Effect of Vermiwash with Biopesticides against Infestation of Pod Borer (Helicoverpa armigera Hub.). International Journal of Zoological Investigations. 2015; 1(1):40-51.

Patil SS, Kengar SB, Sathe TV. New vermiwash model for sustainable agriculture in India. Nature Environment and Pollution Technology. 2007; 6(2):281-284.

Samadhiya H, Dandotiya P, Chaturbedi J, Agarwal AP. Effect of vermiwash on the growth and development of leaves and stem of tomato plants. International Journal of Current Research. 2013; 5(10):3020-3023.

Sathe TV. Vermiculture and Organic Farming. Daya Publisher House, New Delhi, 2004.

Sayyad NR. Utilization of vermiwash potential against insect pests of tomato. International Research Journal of biological Sciences. 2017; 6(1):44-46.

Sharma AK. A Handbook of Organic Farming. Agrobios, Jodhpur, 2004.

Sreenivas Murlidhar S, Rao MS. Vermicompost: A viable component of IPNS in nitrogen nutrition of ridge gourd. Annals of agricultural Research. 2000; 21:108-113.

Sundaravadivelan C, Isaiarasu L, Manimuthu M, Kumar P, Kuberan T, Anburaj J. Impact analysis and confirmative study of physico-chemical, nutritional and biochemical parameters of vermiwash produced from different leaf litters by using two earthworm species. Journal of Agricultural Technology. 2011; 7(5):1443- 1457.

Tharmaraj K, Ganesh P, Kulanjinathan K, Suresh KR, Anandan A. Influence of vermicompost and vermiwash on physico chemical properties of rice cultivated soil. Current Botany. 2011; 2(3):18-21.

Tripathi YC, Hazarika P, Pandey BK. Vermicomposting: An ecofriendly approach to sustainable agriculture. In: Arvind Kumar (eds), Verms and vermitechnology. APH Publishing Corporation, New Delhi, 2005, 23-39.

Palanichamy V, Mitra B, Reddy N, Katiyar M, Rajkumari RB, Ramalingam C, et al. Utilizing Food Waste by Vermicomposting, Extracting Vermiwash, Castings and Increasing Relative Growth of Plants. International Journal of Chemical and Analytical Science. 2011; 2(11):1241-1246.

Verma S, Singh A, Pradhan SS, Singh RK, Singh JP. Bioefficacy of Organic Formulations on Crop Production- A Review. International Journal of Current Microbiology and Applied Sciences. 2017; 6(5):648-665.

Verma S, Singh A, Pradhan SS, Singh RK, Singh JP. Bioefficacy of Organic Formulations on Crop Production-A Review. Int. J Cur. Microbiol. App. Sci. 2017; 6(5):648- 665.

Yadav AK, Kumar K, Singh S, Sharma M. Vermiwash-A liquid biofertilizer. Uttar Pradesh Journal of Zoology. 2005; 25(1):97-99.

Zambare VP, Padul MV, Yadav AA, Shete TB. Vermiwash: Biochemical and Biological approach as eco-friendly soil conditioner. ARPN Journal of Agricultural and Biological Sciences. 2008; 3(4):28-37.

Ansari,A.A. and Ismail,S.A.(2012): Earthworms and Vermiculture Biotechnology, Management of Organic Waste, Dr. Sunil Kumar (Ed.), InTech,Available from: http://www.intechopen.com/ books/management-of-organic-waste/earthworms-and-vermiculture-biotechnology

Ansari,A.A. and Sukhraj,K.(2010): Effect of vermiwash and vermicompost on soil parameters and productivity of okra (Abelmoschus esculentus) in Guyana. Pakistan Journal of Agricultural Research, Vol. 23(3-4): 137-142

Arancon,N.Q., Edwards,C.A, Lee,S., Byrne,R.(2006): Effects of humic acids from vermicomposts on plant growth European Journal of Soil Biology,42:S65–S69

Bharadwaj,A.(2010):Management of Kitchen Waste Material through Vermicomposting, Asian J. Exp. Biol. Sci., Vol. 1 (1):175-177

Basheer, M. and Agrawal, O.P (2013): Management of paper waste by vermicomposting using epigeic earthworm, Eudrilus eugeniae in Gwalior, India Int.J.Curr.Microbiol. App.Sci 2(4): 42-47

Bhawalkar, U.S. (1995): 'Vermiculture Ecotechnology', Bhawalkar Earthworm Research Institute, Pune, India.pp.332

Buckerfield, J.C, Flavel, T.C, Lee, K.E. and Webster, K.A. (1999): Vermi- compost in solid and liquid forms as a plant growth promoter. Pedobiologia, 43(6).753-739.

CRIDA (2009): Vermicompost from Waste; Pub. Of Central Research Institute for Dryland Agriculture; Hyderabad, (Unit of Indian Council for Agricultural Research).

Chaoui,H.I, Zibilske,L.M. and Ohno,T.(2003): Effects of earthworms casts and compost on soil microbial activity and plant nutrient availability. Soil Biology and Bio-chemistry,35(2):295-302.

Chattopadhyay,G.N. (2012): Use of vermicomposting biotechnology for recycling organic wastes in agriculture. International Journal of Recycling of Organic Waste in Agriculture, 1(8):1-6

Darwin,C. (1881): The formation of vegetable mould through the action of worms, with observations on their habitats. Murray, London. pp.326

Dhanalakshmi.V, K. M. Remia, R. Shanmugapriyan and K. Shanthi (2014) Impact of addition of Vermicompost on Vegetable Plant Growth. Int. Res. J. Biological Sci. Vol. 3(12), 56-61

Diaz Cosin, D.J., Novo, M. and Fernandez, R. (2011): Reproduction of Earthworms: Sexual Selection and Parthenogenesis (Chapter 5) In: A. Karaca (ed.), Biology of Earthworms, Soil Biology 24, Springer-Verlag,Berlin Heidelberg

Edwards,C.A. (1988): Breakdown of animal, vegetable and industrial organic waste by earthworms. Agric. Ecosyst. Environ, 24: 21-31.

Edwards,C.A. (1998): Breakdown of animal, vegetable and industrial organic wastes by earthworms. In: Earthworms in waste and environmental management. Edwards, C.A., Neuhauser, F.(Eds). SPB Academic Publishing, The Hague, 21-31.

Edwards,C A. and Arancon, N. (2004): Vermicompost Suppress Plant Pests and Disease Attacks. Rednova News, http://www.rednova.com/display Edwards,C.A., Arancon,N.Q., Emerson, E. and Pulliam,R.(2007): Suppressing plant parasitic nematodes and arthropod pests with vermicompost teas. BioCycle, 48(12):38-39.

Gandhi,M., Sangwan,V., Kapoor,K.K. and Dilbaghi,N. (1997): Composting of household wastes with and without earthworms. Environment and Ecology 15(2):432–434.

Giraddi,R.S. (2003): Method of extraction of earthworm wash: A plant promoter substance.VIIth National Symposium on Soil Biology and Ecology, Bangalore.

Giraddi,R.S., Patil,B.G., Lingaraju,P.S., Umapathy,A.N., Swamy,B.C. and Megalaman,I.R. (2008): Vermitechnology for Successful Management of Muncipal Wastes - A Joint Effort in South India Karnataka J. Agric. Sci., 21(2): 284-286

Govindan,V.S.(1998): Vermiculture and Vermicomposting, Ecotechnology for pollution control and Environmental Management.pp.48-57

Gunasekaran,S. and Desai, S.(1999): Vermicomposting and Spectral Analysis of Vermicompost. Asian.chem.Lett. 3:231-232.

Hashemimajd,K. and Jamaati-e-Somarin, S. (2011): Investigating the effect of iron and zinc enriched vermicompost on growth and nutritional status of peach trees Scientific Research and Essays,Vol. 6(23):5004-5007

Higa,T.(1991): Effective microorganisms: A biotechnology for mankind. pp. 8-14. In: J.F. Parr, S.B.Hornick, and C.E. Whitman (ed.) Proceedings of First International Conference on Kyusei Nature Farming. USDA, Washington.D.C., USA

Ismail,S.A.(1997): Vermicology 'Biology of earthworms' Orient Longman Limited, Chennai, India. ISBN 81-250-10106.

Kale,R.D.(1998):Earthworm: Nature's gift for utilization of organic wastes, 'Earthworm Ecology'(ed)Edwards, C.A. St. Lucie Press New York. pp.355- 376.

Kale,R.D.(1998): Earthworm Cinderella of Organic Farming. Prism Book Pvt Ltd, Bangalore, India.pp.88.

Kale,R.D. and Sunitha,N.S.(1995):Efficiency of earthworm E. eugeniae in converting the solid waste from aromatic oil extraction units into vermicompost, Journal IAEM, Vol. 22, No. 1,267- 269.

Lakshmi,B.L. and Vijayalakshmi,G.S. (2000): Vermicomposting of sugar factory filter press mud using African earthworm species E. eugeniae, Pollution Research,Vol.19, No.3:481-483.

Manaf,L.A, Jusoh, M.L.C, Yusoff,M.K, Hanidza,T., Ismail, T, Harun,R. & Juahir,H. (2009): Influences of Bedding Material in Vermicomposting Process. Journal of Biology ,1(1): 81-91

Murali,M., Bharathiraja,A. and Neelanarayanan, P. (2011): Conversion of Coir Wastes (Cocos nucifera) Into Vermicompost by Utilizing Eudrilus Eugeniae and Its Nutritive Values. Ind. J. Fundamental and Applied Life Sciences, Vol. 1 (3): 80-83

Nagavallemma,K.P., Wani,S.P., Stephane,L., Padmaja,V.V., Vineela,C., Babu Rao, M. and Sahrawat,K.L.(2004):Vermicomposting: Recycling wastes into valuable organic fertilizer. Global Theme on Agrecosystems Report no.8.Patancheru 502 324, Andhra Pradesh, India: International Crops Research Institute for the Semi-Arid Tropics.pp.20.

Nair,J., Sekiozoic,V. Anda,M.(2006): Effect of pre-composting on vermicomposting of kitchen waste Bioresource Technology ; 97: 2091–2095

Padashetty,S. & Jadesh ,M. (2014): Earthworm distribution with special reference to physicochemical parameters in Aland and Chincholi regions, Gulbarga district, Int. J. Res. Engg & Tech., Vol. 2,11:1-6

Parray, M.A, Wani, M.G., Rather, G M. (2014): Stimulatory Influence of Some Additives on Vermicomposting by Eudrilus Eugeniae International Journal of Engineering Science Invention, Vol. 3,7: 01-04

Parthasarathi,K., Ranganathan,L.S., Thirumalai,M., Parameshwaran,P. (1999): Mono and polyculture vermicomposting press mud enhance macronutrients, Asian J. Microbial Biotech & Env. Science, Vol. 1:63-65.

Paul,L.C. and Metzger,J.D. (2005):Impact of vermicompost on vegetable transplant quality;HortScience :40(7):2020-2023

Prabha,M.L, Priya, M.S and Pavithra,R.(2014): Microorganisms in the gut of Earthworm Eudrilus eugeniae, Int.J.Curr.Res.Chem.Pharma.Sci.1(3): 06-09

Pucher,J., Ngoc,T. N., Thi Hanh Yen ,T., Mayrhofer,R., El-Matbouli, M., Focken, U. (2014): Earthworm Meal as Fishmeal Replacement in Plant based Feeds for Common Carp in Semi-intensive Aquaculture in Rural Northern Vietnam, Turk. J. Fish. Aquat. Sci. 14: 557-56

Radhakrishnan,B. and Muraleedharan,N.(2010):A Monograph on Vermiculture and Vermicomposting in tea growing areas. A joint Publication from UPASI Tea Research Foundation, Coimbatore and National Tea Research Foundation, Kolkata, pp.76

Rajasekar,K., Daniel,T. and Karmegam,N. (2012): Microbial Enrichment of Vermicompost. International Scholarly Research Network ISRN Soil Science Vol 2012,Article ID 946079,pp.13

Rekha,G.S., Valivittan, K. and Kaleena, P.K. (2013): Studies on the Influence of Vermicompost and Vermiwash on the Growth and Productivity of Black Gram (Vigna mungo),Adv. in Biol. Res.,7(4):114-121

Ronald,E.G. and Donald, E.D. (1977): Earthworms for Ecology and Profit, Vol.1,'Scientific Earthworm Farming.' Bookworm Publishing Co., Ontario, California.

Saumaya, G., Giraddi, R. S. and Patil, R. H. (2007): Utility of vermiwash for the management of thrips and mites on chilli (Capiscum annum) amended with soil organics. Karnataka J. Agriculture Sciences, 20(3):657- 659.

Seenappa,C., Jagannatha, C. B., Kale, R. D.(1995): Conversion of distillery waste into organic manure using earthworms E. eugeniae, Journal IAEM, Vol. 22, No. 1:244-246

Sharma,S., Pradhan,K., Satya,S., Vasudevan,P. (2005): Potentiality of Earthworms for Waste Management and in Other Uses – A Review; The J. American Science, 1(1):4-16

Sinha,R.K., Agarwal,S., Chauhan,K., Valani,D.(2010):The wonders of earthworms & its vermicompost in farm production: Charles Darwin's 'friends of farmers', with potential to replace destructive chemical fertilizers from agriculture. Agricultural Sciences Vol.1, No.2:76- 94

Singha,R.P., Singha,P., Araujoc,A.S.F., Ibrahim,A.H. and Sulaiman,B.O. (2011): Management of urban solid waste: Vermicomposting a sustainable option Resources, Conservation and Recycling, 55: 719–729

Sreenivasan, E.(2013a):Bioconversion of industrial wood wastes into vermicompost by utilizing African Night Crawlers (Eudrilus eugeniae). Int.J.Adv. Engg.Tech,IV/III/JulSept.2013:19-20

Sreenivasan, E.(2013b):Evaluation of Effective Microorganisms Technology in industrial wood waste management.Int.J.Adv.Engg.Tech,IV/III/Jul-Sept.2013/21-22

Suhane,R.K.(2007):Vermicompost.Publication of Rajendra Agriculture University, Pusa, Bihar, India, pp.88

Suthar,S. (2010): Evidence of plant hormone-like substances in vermiwash: An ecologically safe option of synthetic chemicals for sustainable farming. J. Ecological Engineering, 36(8):1089-1092

Thangavel,P., Balagurunathan,R., Divakaran,J. and Prabhakaran,J.(2003): Effect of vermiwash and vermicast extract on soil nutrient status, growth and yield of paddy. Advances of Plant Sciences, 16(1):187-190.

Tiwari, S.C., Tiwari, B.K. and Mishra, R.R. (1989): Microbial populations, enzyme activities and nitrogen, phosphorus, potassium enrichment in earthworm casts and in surrounding soil of a pineapple plantation. J.Biological Fertility of Soils, 8(1):178-182.

Vasanthi,K., Chairman, K., Michael, J.S., Kalirajan, A. and Singh, A.J.A.R.(2011): Enhancing Bioconversion Efficiency of the Earthworm Eudrilus Eugeniae (Kingberg) by Fortifying the Filtermud Vermibed using an Organic Nutrient On Line Journal of Biological Sciences,11(1):18-22

Zafari and Kianmehr (2012): Pellets from compost, BioResources;7(4): 4704-4714.

INDEX